Achieving Effective Acquisition of Information Technology in the Department of Defense

Committee on Improving Processes and Policies for the Acquisition and Test of Information Technologies in the Department of Defense

Computer Science and Telecommunications Board

Division on Engineering and Physical Sciences

NATIONAL RESEARCH COUNCIL
OF THE NATIONAL ACADEMIES

THE NATIONAL ACADEMIES PRESS
Washington, D.C.
www.nap.edu

THE NATIONAL ACADEMIES PRESS 500 Fifth Street, N.W. Washington, DC 20001

NOTICE: The project that is the subject of this report was approved by the Governing Board of the National Research Council, whose members are drawn from the councils of the National Academy of Sciences, the National Academy of Engineering, and the Institute of Medicine. The members of the committee responsible for the report were chosen for their special competences and with regard for appropriate balance.

This study was supported by Contract No. W911NF-07-C-0115 between the National Academy of Sciences and the Defense Information Systems Agency. Any opinions, findings, conclusions, or recommendations expressed in this publication are those of the author(s) and do not necessarily reflect the views of the organizations or agencies that provided support for the project.

International Standard Book Number-13: 978-0-309-14828-3
International Standard Book Number-10: 0-309-14828-6

Copies of this report are available from the National Academies Press, 500 Fifth Street, N.W., Lockbox 285, Washington, DC 20055; (800) 624-6242 or (202) 334-3313 (in the Washington metropolitan area); Internet, http://www.nap.edu.

Printed in the United States of America

THE NATIONAL ACADEMIES

Advisers to the Nation on Science, Engineering, and Medicine

The **National Academy of Sciences** is a private, nonprofit, self-perpetuating society of distinguished scholars engaged in scientific and engineering research, dedicated to the furtherance of science and technology and to their use for the general welfare. Upon the authority of the charter granted to it by the Congress in 1863, the Academy has a mandate that requires it to advise the federal government on scientific and technical matters. Dr. Ralph J. Cicerone is president of the National Academy of Sciences.

The **National Academy of Engineering** was established in 1964, under the charter of the National Academy of Sciences, as a parallel organization of outstanding engineers. It is autonomous in its administration and in the selection of its members, sharing with the National Academy of Sciences the responsibility for advising the federal government. The National Academy of Engineering also sponsors engineering programs aimed at meeting national needs, encourages education and research, and recognizes the superior achievements of engineers. Dr. Charles M. Vest is president of the National Academy of Engineering.

The **Institute of Medicine** was established in 1970 by the National Academy of Sciences to secure the services of eminent members of appropriate professions in the examination of policy matters pertaining to the health of the public. The Institute acts under the responsibility given to the National Academy of Sciences by its congressional charter to be an adviser to the federal government and, upon its own initiative, to identify issues of medical care, research, and education. Dr. Harvey V. Fineberg is president of the Institute of Medicine.

The **National Research Council** was organized by the National Academy of Sciences in 1916 to associate the broad community of science and technology with the Academy's purposes of furthering knowledge and advising the federal government. Functioning in accordance with general policies determined by the Academy, the Council has become the principal operating agency of both the National Academy of Sciences and the National Academy of Engineering in providing services to the government, the public, and the scientific and engineering communities. The Council is administered jointly by both Academies and the Institute of Medicine. Dr. Ralph J. Cicerone and Dr. Charles M. Vest are chair and vice chair, respectively, of the National Research Council.

www.national-academies.org

COMMITTEE ON IMPROVING PROCESSES AND POLICIES FOR THE ACQUISITION AND TEST OF INFORMATION TECHNOLOGIES IN THE DEPARTMENT OF DEFENSE

WILLIAM H. CAMPBELL, BAE Systems, Inc., *Co-Chair*
DAWN C. MEYERRIECKS,[1] Dawn Meyerriecks, LLC, *Co-Chair*
ROBERT F. BEHLER, MITRE Corporation
PHILIP E. COYLE III, World Security Institute
RENATO A. DiPENTIMA, SRA International (retired)
JOHN M. GILLIGAN, Gilligan Group, Inc.
JOHN GOODENOUGH, Carnegie Mellon University
PAUL J. KERN (NAE),[2] The Cohen Group
H. STEVEN KIMMEL, Alion Science and Technology
DEIDRE A. LEE, Professional Services Council
JOSHUA S. LEVINE, ESP Technologies Corporation
NACHIAPPAN NAGAPPAN, Microsoft Research
FRANK A. PERRY, Science Applications International Corporation
VAHO REBASSOO, The Boeing Company
DANIEL C. STURMAN, Google, Inc.

Staff

JON EISENBERG, Director, Computer Science and Telecommunications Board
KEVIN LEWIS, Senior Program Officer
LYNETTE I. MILLETT, Senior Program Officer
RENEE HAWKINS, Financial and Administrative Manager
VIRGINIA BACON TALATI, Program Associate
MORGAN MOTTO, Program Associate (through April 2009)

---

[1] Dawn Meyerriecks resigned from the committee in September 2009 upon her appointment as Deputy Director of National Intelligence for Acquisition and Technology.
[2] National Academy of Engineering.

COMPUTER SCIENCE AND TELECOMMUNICATIONS BOARD

Preface

The information technology (IT) revolution of the past several decades has dramatically changed the world. The Internet, Web 2.0 technologies, social networking tools, online search engines, text messaging, video teleconferencing, and multimedia-enabled smart-phones with embedded cameras are but a sample of IT-based capabilities that have altered the ways in which people communicate and work.

In the military, IT has enabled profound advances in weapons systems and the management and operation of the defense enterprise. A significant portion of the Department of Defense (DOD) budget is spent on capabilities acquired as commercial IT commodities, developmental IT systems that support a broad range of warfighting and functional applications, and IT components embedded in weapons systems. The ability of the DOD and its industrial partners to harness and apply IT for warfighting, command and control and communications, logistics, and transportation has contributed enormously to fielding the world's best defense force.

But despite the DOD's decades of success in leveraging IT across the defense enterprise, the acquisition of IT systems continues to be burdened with serious problems. Accordingly, the Defense Information Systems Agency (DISA) asked the National Research Council (NRC) to assess the efficacy of the DOD's acquisition and test and evaluation (T&E) processes as applied to IT. In response, the NRC formed the Committee on Improving Processes and Policies for the Acquisition and Test of Information Technologies in the Department of Defense—a group of IT sys-

BOX P.1
Statement of Task

This study will bring together defense and defense industry experts in acquisition and test and evaluation (T&E); commercial software developers; and software engineers, computer scientists, and other academic researchers to assess the efficacy of the DOD acquisition and T&E processes as specifically applied to information technology. Through briefings, site visits, and committee deliberations, the study committee will:

1. Evaluate legislative requirements for acquisition and T&E and the current DOD acquisition process (as defined in the "DOD 5000 series") to determine whether the law and the defined processes permit enough flexibility to rapidly bring capabilities to users;
2. Examine the processes and capabilities of the commercial IT sector to determine whether industry best practices can be adopted by DOD to improve the acquisition, systems engineering, and T&E process;
3. Examine the Department's various concepts for systems engineering and testing in virtual environments, and make recommendations for how to integrate them into a cohesive, efficient, and robust capability;
4. Examine the DOD acquisition environment, including its institutional and cultural dimensions, for barriers that inhibit program managers/acquisition executives from taking advantage of existing flexibility in law and defined processes and recommend solutions; and
5. Make recommendations to responsible agency, executive branch, and legislative officials about how to improve the acquisition, systems engineering, and T&E processes to achieve the Department's net-centric goals.

tems acquisition and T&E experts, commercial software developers; and software engineers, computer scientists, and other academic researchers. The committee was tasked with the following: (1) an evaluation of applicable legislative requirements, (2) an examination of the processes and capabilities of the commercial IT sector, (3) an examination of the DOD's concepts for systems engineering and testing in virtual environments, (4) an examination of the DOD acquisition environment, and (5) the formulation of recommendations on how to improve the acquisition, systems engineering, and T&E processes to achieve the DOD's network-centric goals. (The full statement of task appears in Box P.1.) The tasks were completed in November 2009. This report provides the committee's findings and recommendations, which are based on document reviews, briefings from commercial and military experts in IT systems acquisition, internal deliberations, and the committee members' personal expertise.

Briefings to the committee from staff of the Office of the Secretary of Defense showed that the acquisition of major automated information systems (MAIS) is especially troublesome. This problem has been broadly recognized for years, and there have been many attempts at reform. Nonetheless, today's processes for the acquisition and testing of DOD IT systems often last 5 or more years before delivering solutions to the end users. Given the rapid pace of change in the IT world, it is no wonder that solutions ultimately delivered by DOD IT programs are often considered by end users to be inadequate. Much the same could be said about the historical adoption of IT in the commercial sector, where there have been extraordinary successes and colossal failures. Fortunately, the commercial sector has enjoyed some great successes in recent years by employing agile IT acquisition approaches that can also be leveraged by the DOD.

In examining the current DOD processes for acquiring IT systems and comparing them with the processes adopted by leading-edge firms in the commercial sector, the committee found stark differences. The DOD is hampered by a culture and acquisition-related practices that favor large programs, high-level oversight, and a very deliberate, serial approach to development and testing (the waterfall model). Programs that are expected to deliver complete, nearly perfect solutions and that take years to develop are the norm in the DOD. In contrast, leading-edge commercial firms have adopted agile approaches that focus on delivering smaller increments rapidly and aggregating them over time to meet capability objectives. Moreover, the DOD's process-bound, high-level oversight seems to make demands that cause developers to focus more on process than on product, and end-user participation often is too little and too late. These approaches run counter to agile acquisition practices in which the product is the primary focus, end users are engaged early and often, the oversight of incremental product development is delegated to the lowest practical level, and the program management team has the flexibility to adjust the content of the increments in order to meet delivery schedules.

The committee concluded that the key to resolving the chronic problems with the DOD acquisition of IT systems is for the DOD to adopt a fundamentally different process—one based on the lessons learned in the employment of agile management techniques in the commercial sector. Agile approaches have allowed their adopters to outstrip established industrial giants that were beset with ponderous, process-bound, industrial-age management structures. Agile approaches have succeeded because their adopters recognized the issues that contribute to risks in an IT program and changed their management structures and processes to mitigate the risks. There are clear parallels in the DOD that support making this process change the centerpiece of improving IT acquisition.

For the DOD to succeed in adopting new approaches to IT acquisition, the first step is to acknowledge that simply tailoring the existing processes is not sufficient. DOD acquisition regulations do permit tailoring, but the committee found few examples of the successful application of the current acquisition regulations to IT programs, and those that were successful required herculean efforts or unique circumstances. Changes broader than tailoring are necessary; they must encompass changes to culture, redefinition of the categories of IT systems, and restructured procurement, development, and testing processes as identified in this report. In the aggregate, these changes must realign processes that today are dominated by deliberate approaches designed for the development of large, complex, hardware-dominated weapons systems to processes adapted to the very different world of software-dominated IT systems.

The specific, actionable recommendations made by the committee address the four dimensions of its task discussed above. The body of the report and the appendixes include detailed discussions, rationale, and two proposed new process models for acquiring IT within the DOD. One model is structured for programs focused on the development of new software to provide new functionality or to integrate commercial off-the-shelf (COTS) components (e.g., MAIS programs). The second model is designed for the acquisition of COTS IT hardware, software, or services. Both have parallels in the commercial sector and are especially relevant for acquiring systems that support DOD information enterprise requirements and operate using the DOD IT infrastructure. The changes are not recommended for adoption in acquiring IT components embedded in weapons systems at this time, but the committee believes that as these changes are refined and institutionalized, many will be applicable to IT components of weapons systems as well.

The committee believes that there is an imperative for change, and it strongly urges the DOD to adopt the recommendations offered in this report. Strong support from the highest levels of the DOD will be required to implement changes of the magnitude recommended.

The committee extends its thanks to the individuals listed in Appendix E who briefed the committee. It also thanks Steven Hutchison, DISA Test and Evaluation Executive, for helping to make this study possible, and Dr. Hutchison and Judith Hill for their assistance throughout the course of the study. Finally, the committee extends its thanks and appreciation to Jon Eisenberg, Kevin Lewis, Lynette Millett, and Virginia Bacon Talati of the NRC's Computer Science and Telecommunications Board whose dedicated support made this report possible.

William H. Campbell, *Co-Chair*
Committee on Improving Processes and Policies for the Acquisition and Test of Information Technologies in the Department of Defense

Acknowledgment of Reviewers

This report has been reviewed in draft form by individuals chosen for their diverse perspectives and technical expertise, in accordance with procedures approved by the National Research Council's Report Review Committee. The purpose of this independent review is to provide candid and critical comments that will assist the institution in making its published report as sound as possible and to ensure that the report meets institutional standards for objectivity, evidence, and responsiveness to the study charge. The review comments and draft manuscript remain confidential to protect the integrity of the deliberative process. We wish to thank the following individuals for their review of this report:

Eddie Bair, E. Bair Associates, LLC,
Calvin Carerra, The Carrera Group, Inc.,
Felix Dupré, The Durango Group, LLC,
Bruce A. Finlayson, University of Washington,
Jacques S. Gansler, University of Maryland,
Michael F. Goodchild, University of California, Santa Barbara,
Richard F. Hilliard II, Independent Consultant, Bar Harbor, Maine,
Steven B. Lipner, Microsoft Corporation,
Charles E. McQueary, Independent Consultant, Arlington, Virginia,
Frank Ostroff, Ostroff Consultants Group, LLC,
Stuart H. Starr, National Defense University,
John P. Stenbit, TRW, Inc. (retired),
Kevin J. Sullivan, University of Virginia,

Anthony M. Valletta, SRA International,
George Wauer, Independent Consultant, Centreville, Virginia, and
Peter J. Weinberger, Google, Inc.

Although the reviewers listed above have provided many constructive comments and suggestions, they were not asked to endorse the conclusions or recommendations, nor did they see the final draft of the report before its release. The review of this report was overseen by Butler W. Lampson, Microsoft Corporation. Appointed by the National Research Council, he was responsible for making certain that an independent examination of this report was carried out in accordance with institutional procedures and that all review comments were carefully considered. Responsibility for the final content of this report rests entirely with the authoring committee and the institution.

Contents

Summary and Recommendations

Despite the decades of success that the Department of Defense (DOD) has had in leveraging information technology (IT) across the defense enterprise to build the world's most powerful military force, its acquisition of IT systems continues to be burdened with problems. Briefings to the National Research Council's (NRC's) Committee on Improving Processes and Policies for the Acquisition and Test of Information Technologies in the Department of Defense from the staff of the Office of the Secretary of Defense (OSD) show that the acquisition of major automated information systems (MAIS) remains especially slow. This problem has been broadly recognized for years, and there have been many efforts at reform in the past. Nevertheless, today's processes for the acquisition and testing of the DOD's IT systems all too often deliver solutions that are too late to satisfy the needs of the user community. Current fielding cycles are, at best, two to three times longer than successful commercial equivalents according to presentations to the committee—representing multiyear delays in delivering improved IT systems to warfighters and the organizations that support them. As a result, the DOD is often unable to keep pace with the rate of IT innovation in the commercial marketplace, cannot fully capitalize on IT-based opportunities, and is unable to deliver IT-based capabilities rapidly to meet urgent requirements. Moreover, to the extent that adversaries have access to state-of-the-art IT and can put it to operational use rapidly, the DOD faces risks across the battlespace.

Regarding the adoption and use of IT, the commercial sector has also experienced delays and failures as well as extraordinary successes. Of special interest in the commercial sector, however, is the extent to which new or adaptive corporations using agile approaches for the acquisition and operational deployment of IT capabilities have outstripped industrial giants that were beset with their own ponderous, process-bound, industrial-age management structures.

As called for in the committee's statement of task (see Box P.1 in the preface), this report examines the acquisition, culture, practices, processes, and rules within the DOD as they apply to information technology; assesses whether the DOD could adopt best practices from the commercial sector for IT acquisition, systems engineering, and test and evaluation (T&E); and makes recommendations to improve the speed and effectiveness of IT acquisition programs.

Many studies have recommended reforms to the defense acquisition system—that is, the institutions, processes, and rules that govern the development, procurement, testing, and fielding of new capabilities. A number of these, including a recent study by the Defense Science Board,[1] have focused on IT acquisition and concluded that there is a need for a unique acquisition process for IT. This study reaches the same fundamental conclusion but adds another dimension in its elaboration of differing types of IT systems and offers a suggested acquisition process for each.

SCOPE AND VOCABULARY FOR INFORMATION TECHNOLOGY ACQUISITION PROGRAMS

Information technology is used for a wide variety of purposes in the DOD, a breadth suggested by the definition of "information enterprise" in DOD Directive (DODD) 8000.1.[2] However, the issues, challenges, and potential solutions are not the same for all elements of the information enterprise. Moreover, the present defense systems acquisition vocabulary

---

[1] Defense Science Board, *Report of the Defense Science Board Task Force on Department of Defense Policies and Procedures for the Acquisition of Information Technology,* Office of the Under Secretary of Defense for Acquisition, Technology and Logistics, Washington, D.C., March 2009.

[2] DOD Directive 8000.1 (Management of the Department of Defense Information Enterprise) defines "information enterprise" as follows: "The DoD information resources, assets, and processes required to achieve an information advantage and share information across the Department of Defense and with mission partners. It includes: (a) the information itself and the Department's management over the information life cycle; (b) the processes, including risk management, associated with managing information to accomplish the DoD mission and functions; (c) activities related to designing, building, populating, acquiring, managing, operating, protecting, and defending the information enterprise; and (d) related information resources such as personnel, funds, equipment, and IT, including national security systems."

embodied in the law and regulations governing IT acquisition[3] does not provide a taxonomy of IT program characteristics suitable to allowing program types to be matched with appropriate acquisition approaches. Consequently, the committee created new definitions for a well-defined subset of the information enterprise that has parallels in the commercial sector.

The committee decided to consider IT systems to be just those systems that support the DOD information enterprise," especially those systems expected to run on or interface with existing infrastructure and systems that are user-facing, and limited to those that are delivered through the acquisition process (and not systems "homegrown" in individual commands). (An *IT program* is defined in this report as the process for acquiring an IT system as defined here.) Excluded from this subset are the IT-based components embedded in weapons systems or DOD-specific hardware. The committee believes that as a general rule such embedded components (for example, firmware that controls a missile's flight) are so integral to the weapons system that they are best managed using the same processes as those used for managing the other elements of the weapon. The committee considers it likely that at least some of the conclusions in this report would also apply to other systems with significant IT content—but this report and its recommendations are focused on opportunities to improve the acquisition of IT systems as defined here.

To better align the acquisition approach with technical characteristics and risk, the committee further subdivides IT systems into two categories:

- *Software development and commercial off-the-shelf (COTS) integration (SDCI) programs*—those focused on the development of new software to provide new functionality or focused on the development of software to integrate COTS components, and
- *COTS hardware, software, and services (CHSS) programs*—those focused exclusively on COTS (hardware, software, or services) without modification for DOD purposes (that is, the capabilities being purchased are determined solely by the marketplace and not by the DOD).

By separating programs that involve software development and/or integration from programs that entail simply adopting COTS technology, these categories facilitate the comparison of defense acquisition processes

---

[3] See, for example, DOD Directive 5002.15000 series regulations; Title 10, U.S.C., Chapters 144 ("Major Defense Acquisition" Programs) and 144A ("Major Automated Information Systems Programs"); and DOD Directive 8000.1.

with commercial best practices and the adoption of practices focused on effective and timely leveraging of the commercial IT cycle.

FINDINGS AND RECOMMENDATIONS

IT Acquisition Process and Culture (Chapter 2)

Findings

The acquisition framework of the Department of Defense[4] prescribes elaborate governance mechanisms and cost thresholds that trigger varying levels of oversight and review. In response to acquisition problems from the past, more oversight and governance, some of it excessively process-centric and adversarial, have been added. Collectively, these well-intended changes have made the timely delivery of IT capabilities more difficult. Today's IT acquisition process often focuses on review documents and other process artifacts, and the acquisition culture does not reward early and transparent feedback on capabilities and limitations. Success as determined by process metrics in acquisition does not necessarily align with success metrics based on the timely delivery of end-user capability.

The DOD's perceived need for caution over speed is understandable. Given the criticality and danger of its mission, its worldwide operations and large workforce, and the frequent need for clear, decisive action, the DOD by its nature is an organization with a classic command-and- control culture. However, if current trends continue, it is likely that processes and systems will become even more top-down and centralized, in spite of the DOD's desire to move to an integrated, cross-Service environment with empowered decision making at all levels of command.

DOD systems acquisition policies, expertise, practice, and culture—including those applied to IT systems—reflect the practices, policies, and cultural norms associated with large weapons systems programs. This report does not address the issue of whether the process for weapons systems is appropriate. Other studies and reports have focused on that question—and many have pointed out shortcomings. With respect to IT, however, there is a long-standing reluctance to deviate from standard weapons system acquisition processes, and acquisition personnel are not trained or led to differentiate the unique aspects of IT systems acquisition.

---

[4] As set forth in Title 10, U.S.C., Chapters 144 ("Major Defense Acquisition Programs") and 144A ("Major Automated Information System Programs") and DOD Directive 5002.15000 series regulations.

The application of current, weapons-system-based acquisition processes to IT systems has a number of deleterious effects on the DOD's ability to deliver needed end-user capability, including the following:

- With the exception of hardware and licensed software purchased through vehicles such as the DOD Enterprise Software Initiative contracts, applicable COTS technologies are insufficiently leveraged, excessively tailored, inefficiently tested, and excessively delayed. Many programs have experienced acquisition or integration lead times that significantly exceed the life cycles of the underlying COTS technology. The discrepancy between DOD fielding cycles and COTS life cycles is stark, and measured in years.
- The oversight process focuses too much on shortcomings of COTS products and services and inhibits the timely delivery of meaningful (albeit imperfect) end-user capabilities.
- IT program requirements are often written with overly detailed specifications, take a long time to develop, and are not consistent with the pace of technological change or the rapid delivery of end-user capabilities.
- DOD acquisition, budgeting, and requirements processes, which are designed for large weapons systems acquisition programs, are being inappropriately applied to relatively low-dollar IT programs.
- Dollar thresholds are used to assign the level of oversight for IT programs. These levels are significantly lower than the dollar levels used for determining oversight levels for weapons system programs. This disparity subjects too many IT programs to time-consuming, high-level DOD oversight and prevents the delegation of oversight to lower levels that are more agile.
- The DOD's acquisition training curriculum does not adequately address the special challenges of IT system acquisition or prepare program managers to run IT programs effectively. This shortfall impedes the DOD's ability to assess, adapt, and adopt applicable commercial methods, processes, products, and services.

Recommendations

Recommendation 1. Adopt a new acquisition process tailored for IT systems.

For IT systems, the acquisition processes, which are currently defined by the 5000 series of DOD regulations, should be replaced with a new process designed specifically for the timely and effective acquisition of

IT systems. Elements of this process are detailed in Chapter 3 and in Appendixes B and C.

The supporting recommendations that follow address key areas where an adjustment of acquisition practice and cultural change are needed.

Recommendation 1.1. Emphasize timeliness and end-user mission success in the DOD IT acquisition culture rather than rigid oversight and process compliance.

Cultural aspects of the DOD acquisition process that have an impact on the potential success of IT acquisition efforts include the following: the bias that larger is better, the sense that oversight personnel have no accountability for delaying needed IT capabilities, an emphasis on process risk (executing the acquisition process correctly) rather than on the risk of late delivery of end-user capability, an unwillingness to admit program failure, an emphasis on process over product, a belief that the DOD is genuinely unique, and the belief that what is good for large weapons systems should be good enough for IT systems.

In addition to implementing a new acquisition process for IT capabilities, the DOD should institute a companion program to address the cultural changes required to make the new acquisition process successful. As examples, the DOD should require that all personnel in the oversight process have accountability for the program (that is, for helping the program succeed or helping terminate programs that are fatally flawed or no longer required). The focus should be on providing products with value to end users, not the oversight process per se. It is also important that the oversight culture support small, incremental cycles for IT development and fielding.

Recommendation 1.2. State IT systems requirements as top-level mission expectations (that is, "big-R" requirements) rather than as detailed processes or technical solutions; develop the details ("small-r" requirements) by iterative refinement with users.

Today, there is an extensive requirements-definition period for the typical IT program, resulting in a large volume of requirements documentation that must be formally validated by the Joint Requirements Oversight Council for major joint programs. This requirements-definition and budgeting process encourages the aggregation of requirements into larger programs. In turn, these requirements documents are used to justify program budget requests, and the approval of these requests is a condition for proceeding to successive acquisition program milestones. Because the requirements-definition and approval processes introduce significant delays, requirements documents for IT programs often have become inaccurate descriptions of user needs by the time that funding is obtained and the acquisition process is initiated.

Instead, IT systems requirements should be defined at the mission capability level. Top-level requirements (big-R) should be rapidly developed and validated, and more detailed requirements (small-r) should be developed as an integral part of the acquisition process.[5] Specifically, based on iterative interactions with the actual end users of the IT capability as well as on assessments of available technology, more detailed requirements would be developed for individual programs or projects. As the acquisition effort progresses, feedback from operational users after deployment of increments of capability would be used to feed the continued evolution of (small-r) requirements for the IT acquisition effort. Doing so will permit iterative refinement with users and the best use of commercially available technologies. At the same time, this recommendation should not be construed to suggest that big-R requirements will not change. Feedback on those is also important and should be accommodated.

The committee's recommended process is similar to one recently established by a Joint Capabilities Integration Development System policy. The approach is currently being applied to a small set of programs; the committee encourages the expansion of this concept to encompass all IT programs.

Recommendation 1.3. Leverage flexibilities within IT acquisition funding to achieve speed and agility in the new acquisition process.

The DOD's process for obtaining funding for new acquisition programs typically takes a number of years. In summary, a DOD capability shortfall is linked to a request to Congress for funding that would be provided in a future year. For solutions that will rely on IT, the time frame for seeking funding can be many times longer than the actual time needed to develop or procure the solution. To achieve rapid delivery of IT solutions, a more responsive process is needed for justifying and allocating funding to address capability shortfalls.

In the short term, the DOD should work with Congress to explore how to make use of flexibility consistent with current legal requirements.

---

5 "Big-R" requirements convey a widely recognized purpose, mission, and expected outcome. "Small-r" requirements provide a set of more detailed requirements associated with specific user interfaces and utilities that will evolve within the broader specified architecture as articulated in the initial big-R requirements.

(Even where legislative changes are not needed, flexibility must nonetheless be negotiated with the congressional defense oversight committees.) For example, the DOD has authority to allocate funds for urgent warfighter needs and to reallocate funding after congressional appropriation as a result of changing needs. This authority could be used to begin the development of a needed capability in weeks or months. Also, acquisition funds are sometimes allocated by Congress to a larger mission or program area, or in some cases to a portfolio of projects identified with an area of mission need—for example, to fund software upgrades in a particular mission area or system. Funding could rapidly be allocated to meet demands for IT capabilities within these areas. In both cases, transparent reporting to Congress will be essential to ensure proper oversight and to demonstrate that the flexibility yields a more rapid delivery of capability to the field.

In the longer term, the DOD should work with Congress to establish a new set of funding mechanisms for meeting IT requirements that would align congressional funding with mission or capability areas rather than with individual acquisition programs. Under this concept, Congress would allocate funding to a mission area that would be governed in the DOD with a portfolio-management-like process. In implementing this concept, DOD officials would be responsible for setting priorities and allocating the funding to individual projects following appropriations for a portfolio of mission requirements. This approach would ensure appropriate justification of funding needs tied to mission requirements during budget submission as well as the rapid allocation of appropriated funding consistent with the pace of evolving mission requirements and technology advancements. Currently, the DOD uses a process similar to this concept for funding maintenance upgrades to aircraft avionics software. Likewise, a somewhat similar process is also used for managing IT projects funded through working capital funding processes.

Recommendation 1.4. Provide IT systems acquisition professionals with education in modern IT systems and establish minimum competency standards.

Training a professional acquisition workforce in modern IT program management is critical. IT systems-acquisition professionals must understand key discriminants of the process, be able to evaluate proposals, and make trade-offs. Their training should include requirements specification; development, integration, and test processes; listening to and incorporating "the voice of the user"; and the use of COTS.

Recommendation 1.5. Use pilot programs to institutionalize the new IT acquisition process recommended in this report.

Pilot programs would provide the DOD with the means to apply new acquisition approaches and to capture valuable lessons learned and develop the necessary guidance for applying the knowledge acquired to larger and future programs. The committee believes that the establishment of 10 pilot programs for the rapid acquisition of IT systems capabilities under the alternative IT acquisition process described in this report would greatly help in moving the acquisition process forward. These pilot programs could be the catalyst for institutionalizing a new and updated IT systems acquisition process and provide an opportunity for the DOD to deepen the IT systems acquisition experience of its workforce.

Recommendation 1.6. Propose legislative and regulatory changes (1) to codify a new agile process for acquiring IT systems and (2) to revise dollar thresholds for the oversight of IT systems acquisition in order to foster decentralization.

The committee proposes that two major initiatives be undertaken to update legislation and regulations governing IT acquisition in the DOD. The first initiative would involve the adoption of a new agile process as the default for the acquisition of IT systems. Central to the new process would be the concept of the iterative, incremental acquisition of capabilities that are delivered to end users as successive packages that are aggregated over time into comprehensive capabilities—a concept that is consistent with today's best commercial practices. This concept would apply to the acquiring of custom IT capabilities that are developed or integrated and are deployed on existing IT infrastructures (SDCI systems) as described above. In addition, a companion process would be tailored to the acquisition of off-the-shelf IT capabilities (CHSS systems). Processes for both types of IT acquisition are described in detail in this report.

The second initiative would be to restructure and decentralize the IT acquisition oversight process in order to align it with the fast-paced cycles of agile and rapid acquisition. Today's acquisition oversight process in the DOD is designed for the disciplined management of large, expensive, and complex weapons systems. However, the dollar thresholds for designating oversight levels for IT programs are significantly lower than those used for weapons systems (by a factor of five). Moreover, the current legislation has no provision for MAIS programs to be designated as acquisition category (ACAT) II, which would provide for oversight at the Service or agency level. One approach to solving the problem of highly centralized oversight with its attendant delays would be to use the same dollar thresholds in effect for major defense acquisition programs for the designation of ACAT levels to MAIS programs. This change in thresholds for IT programs would foster decentralization and would better align authority for IT program oversight to the appropriate levels—at OSD, the Services and agencies, and lower echelons.

IT Systems and Software Engineering Processes and Practices (Chapter 3)

Findings

Information technology programs are profoundly affected by the rapid and relentless pace of change in the underlying technologies. Hardware capability per unit expenditure doubles roughly every 18 months. Software capability is driven by the even-faster pace of technology change in the Internet environment and by the elevated level of end-user expectations that this causes. DOD IT systems acquisition programs progress at a markedly slower pace. As a result, the DOD is unable to keep pace with the rate of IT innovation in the commercial marketplace, cannot fully capitalize on IT-based opportunities, is unable to deliver IT-based capabilities rapidly, and, accordingly, will not have the requisite agility.

Historically, system development has followed a "waterfall" process calling for formally documented specification, followed by a request for bids, followed by contracting, delivery, installation, and maintenance. Major elements of the waterfall method, which is document-intensive, attempt to satisfy management's goals for ensuring that projects will successfully meet their objectives, but they end up emphasizing the acquisition process rather than the capability being delivered.

To deliver software capability more rapidly, the commercial world has widely embraced the iterative, incremental development (IID) approach, which addresses two issues of central importance for IT systems—(1) the need for user interaction in setting requirements and (2) complexity.[6] Key attributes of the IID approach include (1) the prominence of the end user's voice, (2) a focus on big-R requirements during early planning, (3) the close integration of developmental and operational T&E into the development cycle, and (4) the breaking down of a project into incremen-

[6] There is another, alternative strategy based on incremental development strategies—evolutionary acquisition, which addresses a concern somewhat different from that addressed by IID, namely, the issue of technology maturation. Under evolutionary acquisition, early increments provide end-user capabilities based on mature technology, and work on later increments is deferred until needed technology has been matured.

tally deliverable parts. The commercial world has widely embraced the IID approach.

Although the DOD's current governance and oversight structure permits tailoring and provides the flexibility needed for a milestone decision authority and program manager to adjust how the process is applied to specific programs, there is no established best practice or accepted template for tailoring. Instead, each decision maker must independently address on an ad hoc basis the misalignment between an oversight system designed for hardware and weapons system development and the very different types of issues that contribute to risk in an IT system program.

With the existence of multiple oversight bodies and large numbers of participants in the program oversight and review process, the current system gives undue leverage to groups that often are not true stakeholders in the process. This can have negative effects, including too many detailed and ad hoc small-r requirements placed on the program, an inability to prioritize requirements effectively, and "corner case" requirements (focused on rare situations that only occur outside normal operating parameters) that can be contradictory or extremely difficult to implement.

Recommendations

Recommendation 2. Adopt an iterative, incremental approach for acquiring information technology systems.

The following supporting recommendations are aimed at enabling more rapid and nimble IT systems acquisition processes and fostering the institutionalizing of IID practices.

Recommendation 2.1. Establish iterative, incremental development (IID) processes based on agile software development and related approaches as the default for IT system development.

Software engineering, although still a young discipline, has advanced substantially over the past four decades, and much is understood about how to structure development efforts to better manage the unique risks inherent in IT system programs and to improve the probability of success. In particular, it is now widely appreciated that the waterfall development process is not appropriate for most software development projects. One emerging approach to IID that has many attractive attributes is agile software development (ASD),[7] which has at its core ongoing, integrated developmental and operational testing and a close coupling between

[7] A. Cockburn, *Agile Software Development*, Addison-Wesley, Boston, Mass., 2001.

those responsible for requirements, for development, and for testing. It emphasizes frequent testing and interaction over formal, up-front documentation. The adoption and implementation of an agile-inspired IID approach do not equate simply to compressing traditional waterfall models into shorter time periods, nor does appending current document-centric oversight processes to a series of release phases equate to the use of ASD.

IID and ASD are proven approaches for building capabilities in line with end-user expectations and needs and for rapidly iterating requirements and solutions based on end-user feedback. Such approaches have been embraced in the commercial world as being highly effective ways to deliver incremental improvements to the field rapidly. Their adoption not only would conform to widely accepted commercial best practices but also would implement more than 20 years of recommendations from forums such as the Defense Science Board. The adoption of IID approaches coupled with a focus on the end-user experience does not mean, however, that other stakeholders and nonfunctional requirements (such as information assurance, reliability, and so on) are unimportant. Historically, other stakeholder voices have dominated the process to the exclusion of the end user. The committee urges a rebalancing and a focus on end-user mission success, because it is the end user who is in the best position to judge whether a capability is useful and should be fielded.

Recommendation 2.2. Allocate top-level DOD mission expectations (i.e., big-R requirements) across increments and use each increment to define and satisfy detailed requirements (i.e., small-r requirements).

Beyond the highest set of (big-R) requirements, there is a more detailed set of (small-r) requirements for such things as specific user interfaces and utilities that will be developed and will evolve within the context of the big-R requirements. Stated another way, users cannot effectively articulate their requirements without interacting with (even partially) fielded systems—in IT systems, one cannot establish what is needed without a sense of what is possible. Requirements such as the expected user interface and user paradigms and integration with other concurrently evolving systems and security practices all dictate that the initial specification be limited to broad system goals and physical operating conditions and that more detailed requirements be evolved in concert with user feedback garnered during incremental development cycles.

Recommendation 2.3. Establish separate and distinct strategies and processes for acquiring custom versus off-the-shelf IT systems.

There are fundamentally different classes of issues involved when dealing with software development and commercial off-the-shelf integration, or SDCI, versus the commercial off-the-shelf hardware, software, and services, or CHSS, components of IT system programs; therefore, different strategies are appropriate when addressing them. In both cases, rapid change in virtually all aspects of the technology—requirements, capabilities, user expectations, usage and development environments, and so on—is a fundamental factor that must be addressed, and IID acquisition strategies are appropriate. Where the requirement includes COTS hardware or licensed software that can be procured at volume discounts through DOD enterprise contract vehicles (such as Enterprise Software Initiative contracts), decentralized procurement should be encouraged. When the DOD awards contracts requiring that such components are to be integrated into an IT system solution, those COTS products should be acquired by the most cost-effective means available. Options should include authorizing the contractor to procure COTS components from established government vehicles such as Enterprise Software Initiative contracts and providing the COTS components as government-furnished equipment to avoid unnecessary handling or procurement fees. This is particularly true for licensed software when existing enterprise contracts permit direct downloading.

Recommendation 2.4. Establish, employ, and report measures of success that emphasize the end-user experience, including timeliness to field.

The committee fully anticipates that any new IT systems acquisition process that is defined and adopted will need to evolve over time. The committee has identified shortcomings in the present system, and recommends a new direction. An appropriate evaluative framework should consider end-user capability, end-user satisfaction, timeliness, quality, operating costs, and acquisition (that is, improvements in the process and overall competitive strategy).

Recommendation 2.5. Provide a stable budget profile across multiple increments for iterative, incremental development of IT programs.

Stable budget profiles for the IT program acquisition cycle are necessary to ensure that end users' requirements can and will be addressed, some at the outset and others in future increments. This will avoid the unintended but real consequence of users attempting to overload requirements into the first capability increment. Multiple time-boxed[8] capability increments will fit within each Planning, Programming, Budgeting, and Execution budget cycle. The confidence of key stakeholders, including users, will increase, as will the transparency of overall program execution status.

IT System Testing (Chapter 4)

Findings

Today, testing discipline is integrated too late and serially into DOD IT systems acquisition practices. Indeed, programs generally defer the testing of IT systems in realistic operational environments until the mandated operational test. Without regular feedback from a user perspective on IT system development, program managers and milestone decision authorities lack critical information for managing and supporting programs, and other key stakeholders lack the knowledge that can build their confidence. This approach stands in contrast to best commercial practice, whereby user testing is performed early and often to ensure that user feedback guides all stages of system development. Because DOD end-user engagement is nonexistent or limited, the test community is unfairly tasked with representing the perspective and needs of end users and is often engaged too late to have a substantive impact on requirements, system architecture, or system functionality. Finally, the acquisition community has been reluctant to embrace virtualized testing and has been overtly precluded from reusing or accessing operationally relevant test data and environments.

Recommendations

Recommendation 3. Perform continuous testing, with early involvement from end users, in acquiring DOD information technology systems.

---

[8] *Time boxing* refers to a deadline-driven approach to system development in which work items may slip from one iteration to the next, but iterations are completed according to schedule, thus affording the opportunity to quickly identify erroneous estimates of the time required to complete deliverables and ensuring continuous user input regarding priorities.

The following supporting recommendations would establish testing and evaluation policies and practices in support of the committee's proposed DOD IT systems acquisition process.

Recommendation 3.1. Adopt continuous testing in DOD IT systems development, and insist on the use of metrics, especially emphasizing measures of end-user satisfaction.

Continuous user testing is an integral part of successful IT systems development and deployment. An *acceptance team* emphasizing the perspective of the end user and composed of operational end users, development testing and operational testing stakeholders, security certification and accreditation stakeholders, and interoperability stakeholders should be continuously integrated into the development process. Such integration is critical to ensuring both that the big-R requirements are the right ones and that the system is meeting both big-R and small-r requirements. To facilitate measuring success, a process involving a robust set of metrics should be in place through an IT system's life cycle, from development to ultimate decommissioning. A process that collects, aggregates, and analyzes metrics on end-user service consumption and experiences will provide visibility into what features end users are actually using. Successful commercial suppliers of IT systems place considerable emphasis on metrics that capture actual end-user behavior measured by online interactions with deployed IT systems.

Recommendation 3.2. Emphasize the needs of end users by having the acceptance team play a lead role in recommending deployment decisions.

Continuous user input is critical to the successful development of new IT systems. There are numerous examples of new IT systems meeting every formal aspect of their specifications but not delivering the expected value to their users; conversely, a system that may not meet all aspects of a specification may already be an improvement over existing systems. With these situations in mind, the acceptance team should play a leading role in recommending deployment decisions. This would prevent the deployment of systems that are not adding value (independent of their readiness on paper per the specification), and it would also ensure more rapid deployment of systems that can improve the effectiveness of DOD users (especially the warfighter). Deployments may take the form of small trials, large-scale "beta" programs, or even full production deployments.

Recommendation 3.3. Test with users in their actual work or field environment (sometimes referred to as a beta deployment).

It is essential to put capabilities into the hands of users early and to measure performance and otherwise obtain feedback. Engaging expected users of the system early enough makes it possible to provide meaningful feedback to developers and influence the small-r requirements. One especially useful approach is to engage users through pilot projects deployed in the field, possibly operating in parallel with production systems.

Recommendation 3.4. Accept certification and functional IT system component test results across organizational boundaries.

The DOD has broadly adopted a set of networking capabilities that are integral to every IT system. As a matter of DOD policy, such capabilities should not be required to undergo separate revalidation or recertification during operational testing. Examples include such capabilities as Internet Protocol and Domain Name System software modules. As more of the technology stack becomes commoditized or provided as a service, the set of associated capabilities not requiring revalidation should likewise grow. Current examples include public key infrastructure and on-demand computing and storage. The same policy should apply to DOD-deployed and DOD-certified software modules that are reused across DOD programs. The testing of the composite system product that is evaluated in a realistic environment for operational effectiveness and suitability is sufficient.

Recommendation 3.5. Use virtual test environments to support both continuous feedback and certification of IT program increments.

A variety of opportunities to establish virtual test environments already exist, as identified in Chapter 4. The definition of test environment could, over time, be significantly extended to include the use of beta testing in actual operational environments, the use of commensurate commercial data and/or operating environments, and the extension of virtual test environments to include operations monitoring as a source of continuous test feedback. The DOD should codify this approach and encourage the use of virtual environments. Operational realism is important but must be balanced against pragmatic considerations.

CONCLUSION

The current DOD approach to IT acquisition has not been broadly successful in delivering needed capability in a timely manner. To lever-

1

Introduction

DEFINITIONS OF THE TERM "IT SYSTEM"

The statement of task for this study calls for an examination of the acquisition and the test and evaluation (T&E) processes as specifically applied to information technology (IT) in the Department of Defense (DOD). At the outset of the study, the Committee on Improving Processes and Policies for the Acquisition and Test of Information Technologies in the Department of Defense discovered that the term "IT system" was used in different ways by briefers to the committee as well as among members of the committee itself. Further investigation showed that the DOD provides no specific definition of "IT system" per se (see Box 1.1 for relevant examples). For purposes of this study, the committee decided to consider IT systems to be just those systems that support the DOD "information enterprise" (see definition in Box 1.1), but excluding IT embedded in weapons systems and in DOD-unique hardware. In particular, the term as used by the committee signifies systems expected to run on or interface with existing infrastructure and systems that are user-facing; moreover, "IT system" as used by the committee means systems that are delivered through the acquisition process (and not systems "homegrown" in individual commands).

The committee subdivided IT systems as specified above into two categories that differ in terms of development requirements, technical characteristics, and risk:

BOX 1.1
Definitions Related to the Term "IT System" in Department of Defense Directives

Department of Defense Instruction (DODI) 5000.2, published most recently in December 2008, defines the authoritative DOD acquisition process. The terms "IT system," "information technology system," and "information system" are not explicitly defined in DODI 5000.2, although the term "IT system" is used in several places, as is the term "information system." An "automated information system (AIS)" is defined as follows:

> A system of computer hardware, computer software, data or telecommunications that performs functions such as collecting, processing, storing, transmitting, and displaying information. Excluded are computer resources, both hardware and software, that are:
> a. an integral part of a weapon or weapon system;
> b. used for highly sensitive classified programs (as determined by the Secretary of Defense);
> c. used for other highly sensitive information technology programs (as determined by the ASD(NII)/DOD CIO) [Assistant Secretary of Defense for Networks and Information Integration/DOD Chief Information Officer]; or
> d. determined by the USD(AT&L) [Under Secretary of Defense for Acquisition, Technology and Logistics] or designee to be better overseen as a non-AIS program (e.g., a program with a low ratio of RDT&E [research, development, test, and evaluation] funding to total program acquisition costs or that requires significant hardware development).[1]

This definition focuses on characteristics relevant to the matter of who manages acquisition oversight for various types of programs based on the application, funding, or sensitivity of the program.

DOD Directive (DODD) 8000 specifies oversight responsibilities for DOD information-management activities and supporting information technology, implementing provisions of the Information Technology Management Reform Act of 1996 (part of the National Defense Authorization Act for Fiscal Year 1996, Public Law 104-106). "Information technology" is defined in the directive as follows:

- *Software development and commercial off-the-shelf integration (SDCI) programs*—those that focus on the development of new software to provide new functionality or on the development of software to integrate commercial off-the-shelf (COTS) components, and
- *COTS hardware, software, and services (CHSS) programs*—those that are focused exclusively on COTS hardware, software, or services without modification for DOD purposes (that is, the capabilities being purchased are determined solely by the marketplace and not by the DOD).

> Any equipment or interconnected system or subsystem of equipment, used in the automatic acquisition, storage, manipulation, management, movement, control, display, switching, interchange, transmission, or reception of data or information by the executive agency, if the equipment is used by the executive agency directly or is used by a contractor under a contract with the executive agency that requires the use of that equipment; or of that equipment to a significant extent in the performance of a service or the furnishing of a product. Information technology includes computers, ancillary equipment, software, firmware and similar procedures, services (including support services), and related resources; but does not include any equipment acquired by a Federal contractor incidental to a Federal contract.[2]

This definition of information technology has, unfortunately, been too often interpreted as communications hardware-focused, although its scope is clearly broader.

As a result, the committee chose as its point of departure for this study the definition of the "DOD information enterprise" provided in the glossary of DODD 8000.1:

> Department of Defense Information Enterprise. The DOD information resources, assets, and processes required to achieve an information advantage and share information across the Department of Defense and with mission partners. It includes: (a) the information itself and the Department's management over the information life cycle; (b) the processes, including risk management, associated with managing information to accomplish the DOD mission and functions; (c) activities related to designing, building, populating, acquiring, managing, operating, protecting, and defending the information enterprise; and (d) related information resources such as personnel, funds, equipment, and IT, including national security systems.[3]

---

[1] DOD Instruction 5000.2, "Operation of the Defense Acquisition System," 2008, p. 33.

[2] DOD Directive 8000.1, "Management of the Department of Defense Information Enterprise," 2009, p. 11.

[3] DOD Directive 8000.1, 2009, p. 10.

EFFECTIVE APPROACHES TO INFORMATION TECHNOLOGY IN THE COMMERCIAL SECTOR

The information age has ushered in an era of personalized products and services built on standard, massively replicable platforms—a powerful combination of centrally supported IT and end-user-driven IT (which generally relies on centrally managed IT to provide at least some of the underlying computing, storage, and communications capabilities). The result has been an ever-increasing empowerment of individuals and organizations, giving them the ability to innovate their technical capabilities, their business processes, and their own product and service offerings. Accompanying this empowerment has been a rising set of expectations for performance of the information technology foundations through which these expectations are met. Hence the environment for delivering capability has become increasingly competitive, with emergent, tailored solutions for certain kinds of problems realized in days and months, sometimes by the customers themselves.

How are commercial IT market leaders managing these demands? They are doing so by instituting standardization and discipline at the heart of their respective IT enterprises while enabling agile, customer-led innovation at the edge of these enterprises. Most large IT providers have developed highly reliable, available, and scalable computing environments as the backbone of their product and service offerings. Consider search engines, commodity trading platforms, online auctions, and online marketplaces. All of these are based on commodity hardware and software that have been integrated to provide uninterrupted, extensible computing power, in many cases around the globe. These platforms are defined and their interfaces are exposed, at least internally, with an emphasis on interface stability and longevity.[1,2] In some cases, the platform interfaces are exposed and accessed externally.[3,4]

Exposed, stable interfaces enable customers to apply computing power in new and unanticipated ways without compromising configuration control by the service provider or hindering the overall customer experience. By exposing robust interface points, customers can elect (or build) their own uniquely tailored experiences, thereby enjoying high satisfaction themselves and providing a reliable business base for the supplier.

The perception—and sometimes the reality—is that customer-led innovation is a "free-for-all" at the edge. Indeed, in many cases, consumer-facing providers cannot—or do not seek to—control the edge because their market is so diverse. However, this is *not* the general case for most enterprises. Many companies are successful at actively pursuing customer-led innovation as a principal means of driving the company's evolution while doing so in a methodical, managed way. Integrating

[1] Luiz Andre Barroso, J. Dean, and U. Holzle, "Web Search for a Planet: The Google Cluster Architecture," *IEEE Micro* 23(1):22-28, April/May 2003.

[2] Anand Gangadharan, "eBay Platform Roadmap," eBay Devcon 2009, June 2009, San Jose, Calif.

[3] Association for Computing Machinery, "A Conversation with Werner Vogels, Learning from the Amazon Technology Platform," *ACM Queue* 4(4):14-17, May 2006.

[4] Tom Killalea, "Building Scalable Web Services: Build Only What You Really Need," *ACM Queue* 6(6):10-13, October 2008.

the customer overtly into the product or service evolution is viewed as essential to success.[5,6,7] At the same time, customer-led innovation has not resulted in delivery that is completely customer-driven: commercial developers still have their own rhythms of delivery of features, releases that customers must wait for.

For both centrally defined and edge-defined IT, many successful commercial IT suppliers have organized development around two key principles: (1) portfolio management[8] and (2) development by small teams employing agile software-development methods.[9,10] *Portfolio management* is a formal process whereby limited resources are strategically allocated to a subset of possible projects. Project risk, overall objectives, costs, benefits, and project interdependencies are all weighed, and a corporate-level decision is rendered on strategic investments. Implemented properly, portfolio management is an agile management tool that can accept current, real-world data and quickly evaluate and recommend changes to the portfolio.

The use of small teams employing agile methods has many advantages. Among them is the minimal enterprise expense that is incurred prior to the engagement of the first users and to all subsequent releases until a business base is established. If a product or service fails to meet business objectives at any point in its evolution, it can be canceled or redirected, at relatively low cost.[11] The agile approach is one specific approach to software development within a larger category known as iterative, incremental development (IID). A survey article[12] on the history of IID chronicles a long succession of major technology programs that have successfully used IID, including the X-15 hypersonic aircraft program and the application of IID methods to software projects on NASA's Project Mercury. By the 1970s, IID was more widely applied to major software projects at selected major prime government contractors, including TRW and

---

[5] Nanette Byrnes, "Xerox Refocuses on Its Customers," *Business Week*, April 18, 2007.

[6] "Lego Mindstorms Advanced User Tools." Available at http://mindstorms.lego.com/Overview/NXTreme.aspx; accessed June, 26, 2009.

[7] "National Instruments' LabView." Available at http://zone.ni.com/dzhp/app/main; accessed June 26, 2009.

[8] M.W. Dickinson, A.C. Thornton, and S. Graves, "Technology Portfolio Management: Optimizing Interdependent Projects over Multiple Time Periods," *IEEE Transactions on Engineering Management* 48(4):518-527, November 2001.

[9] Ade Miller and Eric Carter, "Agility and the Inconceivably Large," pp. 304-308 in *Proceedings of the Agile 2007*, IEEE Computer Society, Washington, D.C., 2007.

[10] Association for Computing Machinery, "A Conversation with Werner Vogels," 2006.

[11] Lan Cao and Balasubramaniam Ramesh, "Agile Requirements Engineering Practices: An Empirical Study," *IEEE Software* 25(1):60-37, January/February 2008.

[12] Craig Larman and V.R. Basili, "Iterative and Incremental Development: A Brief History," *IEEE Computer* 36(6): 47-56, June 2003.

IBM. The 1980s and 1990s saw significant evolution in IID approaches, and in 2001 the first text on the subject, *Agile Software Development*, by Alistair Cockburn, was published.[13] A more in-depth discussion of IID is provided in Chapter 3 of this report. This chronology situates agile and related approaches within a broader context and also demonstrates that IID has a long history of being applied successfully for different types and scales of problems both in the DOD and in the commercial sector.

THE DEFENSE ACQUISITION SYSTEM

The complex Defense Acquisition System (DAS) has three major components, defined as follows:

- The *Joint Capabilities Integration and Development System* (JCIDS) is aimed at identifying, assessing, and prioritizing joint military capability needs. The Joint Staff and the Joint Requirements Oversight Council champion it.[14]
- The *Planning, Programming, Budgeting and Execution System* (PPBES) allocates resources to capabilities deemed necessary to accomplish the DOD's missions. The Under Secretary of Defense, Comptroller champions it.[15]
- The *Defense Acquisition Management System* (DAMS) establishes the "management framework for translating capability needs and technology opportunities, based on approved capability needs, into stable, affordable, and well-managed acquisition programs that include weapon systems, services, and automated information systems." The Under Secretary of Defense for Acquisition, Technology and Logistics (USD AT&L) champions the DAMS.[16]

Each of these components is discussed in more detail in Appendix A.

The inherent difficulties in synchronizing these three DAS components have implications for all types of acquisition programs, including those delivering IT systems. The January 2006 report of the Defense Acquisition Performance Assessment (DAPA) project concluded that "the budget, acquisition and requirements processes [of the Department of Defense] are not connected organizationally at any level below the Dep-

[13] A. Cockburn, *Agile Software Development*, Addison-Wesley, Boston, Mass., 2001.

[14] Defense Acquisition University, JCIDS Definition. Available at http://www1.dau.mil/; accessed June 2009.

[15] Defense Acquisition Guidebook Section 1.2. December 2004. Available at https://akss.dau.mil/dag/guidebook/IG-c1.2.asp; accessed June 2009.

[16] DOD Instruction 5000.2, "Operation of the Defense Acquisition System," 2008, Paragraph 1.b.

uty Secretary of Defense."[17] The DAPA panel specifically considered the impact of this disconnect on DOD software-related programs and made a number of recommendations aimed at addressing the problems.

The present committee's report is focused largely on the DAMS component of the DAS. The PPBES is a well-established process, and its demands are largely predictable. The JCIDS requirements are sufficiently general to provide the necessary flexibility, and are integrated with the existing DAMS. The committee believes that the present Defense Acquisition Management System constitutes a significant challenge to the successful acquisition of IT programs and that changing it represents a promising opportunity to improve the performance of these programs. Moreover, the committee believes that these changes can be successfully integrated with the other existing components of the DAS. The DAMS thus constitutes the focus of this report, although the changes proposed by the committee may also have implications for the JCIDS and PPBES components.

RESULTS OF CURRENT ACQUISITION PROCESSES AND PRACTICES FOR INFORMATION TECHNOLOGY SYSTEMS

The committee received a briefing from the Office of the Assistant Secretary of Defense (Networks and Information Integration) (OASD [NII]) regarding the time that a set of major automated information system (MAIS) programs took to progress through the DOD acquisition system. The set was composed of 23 MAIS programs (3 of which were labeled as extensions of existing programs) that were initiated in fiscal year (FY) 1997 or later and that were completed or discontinued by early 2009. The presentation provided summary charts, and the OASD (NII) later provided the committee with the underlying data.[18] This data set gives the dates on which each program started and completed the following phases in the acquisition cycle: the analysis of alternatives (AoA), the economic analysis (EA), engineering and manufacturing development (which begins following Milestone B [MS B]), and the achievement of initial operating capability (IOC). Some programs started a phase without completing previous phases, and some programs completed a phase without continuing to the next phase.

In the figures and table in this chapter, those programs that entered

---

[17] Assessment Panel of the Defense Acquisition Performance Assessment Project, *Defense Acquisition Performance Assessment Report*, Department of Defense, Washington, D.C., January 2006.

[18] Timothy J. Harp, Deputy Assistant Secretary of Defense (C3ISR & IT Acquisition), "Information Technology Acquisition," presentation to the committee, Washington, D.C., February 25, 2009; and Timothy J. Harp, personal communication to the committee.

the acquisition process at AoA are labeled A to H. (These labels are used rather than the program names because the objective of the analysis was to establish time lines rather than to examine issues associated with individual programs.) The programs that entered EA without first completing an AoA are labeled AA to DD. The programs that started at MS B are labeled AAA to HHH.

Eight programs started the acquisition process at the AoA phase (pre-Milestone B). Figure 1.1 indicates the time in months for each program to complete its AoA. The average time for these programs to complete their AoA was 11 months; the median was 13 months. Five of these programs (those labeled "A", "B," "C," "D," and "H") went beyond AoA completion.

Nine programs in this data set completed their economic analysis (Figure 1.2). Five of these nine were continuations of efforts shown in Figure 1.1. The five programs in common in Figures 1.1 and 1.2 took an average of 28 months and a median of 30 months to complete both phases—roughly 2½ years. The remaining four programs reflected in Figure 1.2 entered EA without first completing an AoA. Overall the aver-

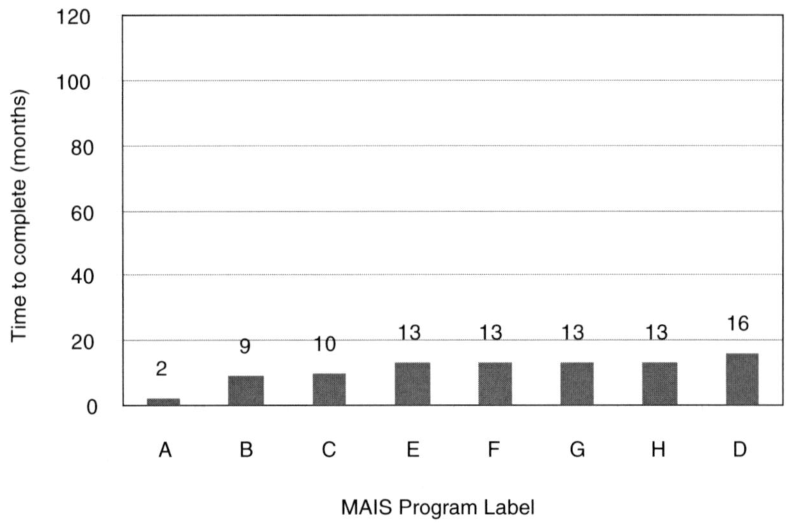

FIGURE 1.1 Time taken to complete the analysis of alternatives (AoA) for the eight major automated information system (MAIS) programs that started the acquisition process at the AoA phase. NOTE: See the accompanying text for an explanation of the program labels. SOURCE: Compiled by the committee from data provided by the Department of Defense for 23 MAIS programs initiated in FY 1997 or later and completed or discontinued by early 2009.

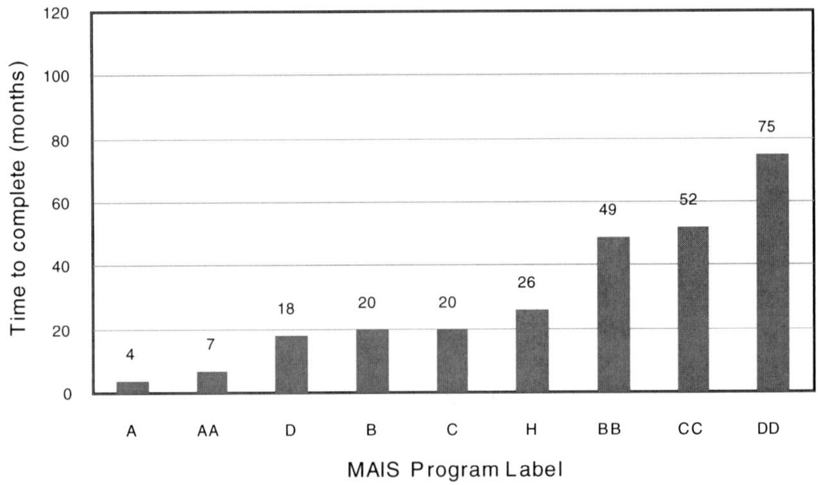

FIGURE 1.2 Time taken to complete the economic analysis phase (AoA completion to Milestone B) for major automated information system (MAIS) programs during FY 1997 to early 2009. NOTE: See the accompanying text for an explanation of the program labels. SOURCE: Compiled by the committee from data provided by the Department of Defense.

age time for programs in this data set to complete the EA was 30 months; the median was 20 months.

Figure 1.3 shows the time that it took for 13 programs to go from Milestone B to a successful initial operating capability. Most of these programs entered the acquisition process at Milestone B. Two of these programs (labeled "A" and "D") completed all three phases of the acquisition process and are represented in all three figures. These programs took a total of 58 months (for "A") and 64 months (for "D") to reach IOC. Overall the average time for programs in this data set to go from Milestone B to IOC was 53 months; the median was 43 months.

Table 1.1 shows the average and median times required across all three acquisition phases to reach IOC. Although it is not mathematically accurate simply to add the averages or medians shown here, these statistics suggest that 6 to 8 years could be required to complete the entire acquisition process and reach IOC. Note that oversight attention is generally believed to have increased over the period of time represented in this data set and analysis, suggesting that the time to IOC may be even longer for more recent programs (and for programs in the future) than these averages suggest.

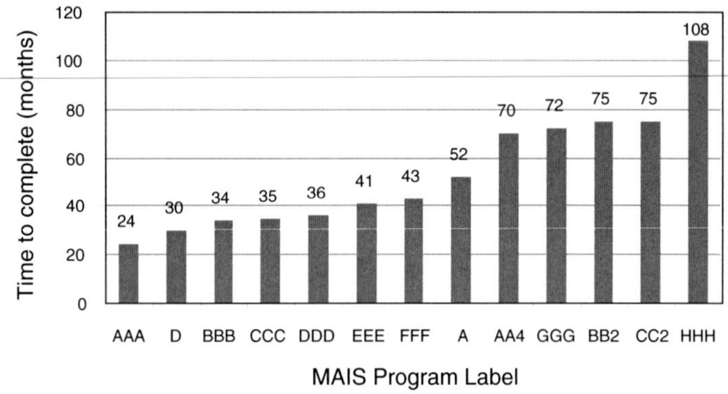

FIGURE 1.3 Time taken from Milestone B to initial operating capability for major automated information system (MAIS) programs during FY 1997 to early 2009. NOTE: See the accompanying text for an explanation of the program labels. SOURCE: Compiled by the committee from data provided by the Department of Defense.

TABLE 1.1 Average and Median Times Taken by Major Automated Information System Programs in Acquisition Process Phases Leading to Initial Operating Capability

Phase	Average (in months)	Median (in months)
AoA completion	11	13
AoA to MS B	30	20
MS B to IOC	53	43

NOTE: See accompanying text for a description of the MAIS programs in the data set; see also Figures 1.1, 1.2, and 1.3. AoA, analysis of alternatives; MS B, Milestone B; IOC, initial operating capability.

One reason for these very long time lines is the burden imposed by the oversight process—the time associated with preparing documentation, scheduling review meetings, and so forth. To illustrate this point, the Business Transformation Agency (BTA)) constructed a graph—referred to as "The Big Ugly," and based on one originally constructed by the U.S. Air Force—that shows all of the reviews and documents required to field a program. The BTA also considered the specific case of adding a 200-line program to a business system and projected that it would take more than

$1 million and 2 years just for the DOD 5000 acquisition reviews and documentation.[19]

SCOPE AND CONTEXT OF THIS REPORT

Over the years, numerous reports have made recommendations aimed at reforming defense acquisition. Indeed, multiple recent reports have tackled the question of IT acquisition specifically and have come to conclusions similar to those reached in this report. The committee believes that this general consensus buttresses the points made here. It is not the committee's purpose, however, to comment specifically on other reports. One distinctive contribution of this report is its discussion of different classes of IT and how such differences merit different acquisition approaches.

The rest of the report examines in more detail the implications of current DOD IT acquisition processes and the committee's rationales and recommended changes. Chapter 2 explores the cultural backdrop of the defense IT acquisition community and its effects on how IT systems are procured. Chapter 3 examines software and systems engineering practices and proposes a revised acquisition-management approach for IT systems. Chapter 4 considers testing and how the testing and evaluation of IT systems within the acquisition process might be made more effective. Appendix A provides a brief overview of the defense acquisition system for IT, Appendixes B and C respectively provide details of the recommended acquisition process for SDCI and CHSS programs, Appendix D gives examples of programs that have succeeded with nontraditional oversight, Appendix E lists briefings provided to the committee, and Appendix F provides biosketches of the committee members and staff. The acronyms used in the report are defined in Appendix G.

[19] Information provided to the committee by Keith Seaman, Acting Director, BTA Component Acquisition Executive, February 2009.

2

The Acquisition Process and Culture

INTRODUCTION

Until 1996, there was a separate set of processes and policies for the acquisition of Department of Defense (DOD) information technology (IT) systems, as called for in the Brooks Act of 1965 (Public Law 89-306). In 1996, the IT and non-IT policies were merged, under the rationale that the requirements of the Brooks Act and the associated DOD 8000 processes had made the acquisition process too cumbersome and slow. IT programs thereafter fell under a single acquisition process specified in the DOD 5000 series regulations, which were intended to provide a more flexible and nimble framework for all types of programs. Shortly after the IT programs were consolidated under DOD 5000, there was an emphasis on tailoring the oversight and documentation requirements of DOD 5000 to better suit the needs of IT programs. There have also been repeated efforts, most recently in December 2008, to reform the process defined by the DOD 5000 series in order to address persistent challenges in the acquisition system.

The DOD has struggled to provide affordable and effective military capabilities through the defense acquisition system (and not just for IT systems). Repeated attempts have been made to reform the acquisition processes, largely based on the experience with weapons systems, with particular attention to highly visible problems in very large programs. Within the Department of Defense, the top 10 acquisition programs account for about 80 percent of the acquisition budget.[1] These programs are large weapons systems programs such as those for the Joint Strike Fighter (F-35), the Future Combat Systems, the Ballistic Missile Defense System, and the SSN 774-class fast-attack submarine. Perceived weaknesses in the DOD acquisition process have included the following: an inadequate assessment of technological maturity before beginning system development; insufficient government reviews and oversight during the multiyear design and development phases; and inadequate preparation for and execution of operational testing.[2,3] In revising the DOD acquisition policies and procedures over the years, the DOD has attempted to address these perceived weaknesses by including more process steps and additional reviews.

Many argue that these reforms, especially the introduction of more oversight, have not improved—and may well have further burdened—the acquisition system. In particular, these changes, aimed primarily at challenges related to large, weapons system programs, have had noteworthy adverse implications for IT programs. IT program managers who provided briefings to the committee during the course of this study indicated that DOD 5000 processes dramatically increase the time to deliver solutions, especially those available as commercial off-the-shelf (COTS) solutions. In addition, the DOD 5000 processes result in the creation of larger formal acquisition programs that, by their very nature, increase documentation requirements and the associated sizes of the support teams.

A recent review conducted by the DOD Obama-Biden Presidential Transition Team noted unanimous agreement among the chief information officers (CIOs) of the DOD and the Military Services that the ability of the DOD acquisition process to deliver needed IT systems was "fundamentally broken." The CIOs cited the inability of the DOD acquisition system to field systems based on commercial technology while it was still state of the art. With commercial IT technologies evolving on 18-month cycles, taking 6 to 8 years for a large IT program to field initial operating capabilities (IOCs; see Chapter 1) is a clear indicator that the defense acquisition processes are not matched to the fundamental characteristics of commercial technology. The CIOs suggested an urgent need to be

---

[1] Nancy Spruill, Director, Acquisition Resources and Analysis, Office of the Under Secretary of Defense for Acquisition, Technology and Logistics, Department of Defense, "Defense Acquisition from a Management Perspective for the NRC Study on IT Acquisition," presentation to the committee, September 2008, Washington, D.C.

[2] Moshe Schwartz, *Defense Acquisitions: Overview, Issues and Options for Congress*, Congressional Research Service, Washington, D.C., June 2008.

[3] Assessment Panel of the Defense Acquisition Performance Assessment Project, *Defense Acquisition Performance Assessment Report*, Department of Defense, Washington, D.C., January 2006.

able to field rapidly to military personnel the same commercial technology available to private-sector users and, in some instances, adversaries. Moreover, the CIOs cited the enormous cost of acquisition oversight processes and the repeated failure of large DOD IT acquisition programs to deliver needed user capabilities. They also noted the need to move toward smaller IT acquisition projects to reduce the risk of failure and to accelerate the fielding of mission-critical capabilities.[4]

The rest of this chapter describes cultural and organizational issues that pose challenges to effective and efficient IT acquisition in the DOD. It closes with a discussion of the importance of measures of success for IT programs together with a brief set of potential metrics.

DIFFERENCES BETWEEN INFORMATION TECHNOLOGY SYSTEMS AND WEAPONS SYSTEMS ARE NOT REFLECTED IN CURRENT PROCESS

Information technology programs have a number of characteristics that distinguish them from defense weapons systems programs. During this study the committee frequently heard the phrase "IT is fundamentally different." These differences are seen as the root cause of many of the problems noted above by the DOD CIO community. As an example, one of the characteristics of most IT programs is their dependence on COTS technologies—thus technology development and many other pre-Milestone B activities specified in the DOD 5000 series are unnecessary. Unfortunately, the tendency in the DOD has been to force-fit DOD 5000 processes, including pre-Milestone B activities—onto COTS-based IT programs and to require that the programs conduct technology demonstrations with mature (or legacy) commercial technology. The program manager (PM) is forced to demonstrate mature and proven technology—and is then forced to carry out the fielding using exactly that same technology baseline, even though, because of the slow pace of the process, the fielding is likely to occur well after the demonstrated software is commercially retired.

As another example of how IT programs are different from weapons systems programs, consider the use and timing of operational test and evaluation (OT&E) activities. The OT&E of weapons systems programs is used as a major risk-reduction activity. For weapons systems, OT&E is conducted by specially trained test professionals charged with determining if the system should go into full-scale production, which represents the majority of the total program cost. For many IT systems, however, the bulk of the program costs have already been spent by the time the opera-

[4] John Gilligan, Gilligan Group, Inc., personal communication to committee, August 2009.

tional testing phase is reached.[5] Therefore, if testing is to inform spending decisions, it must be performed much earlier in the development cycle, and risks must be addressed at that stage.

Moreover, operational testing for an IT system is best done not by professional testers but by the actual users of the system in a limited operational environment. A key question for IT operational testing is whether the system represents an operational improvement over existing capabilities that users are willing to accept in the field—not whether the system should go into full-scale production.

Weapons systems and IT systems are also different in terms of the applicable technology cycles. The technology cycle for IT systems is much more rapid than that for most weapons systems technologies. Although new weapons systems technologies (for example, stealth, engines, and explosive technologies) can take years to develop, there is roughly an 18-month cycle for new IT technologies, which are driven largely by the commercial sector. With the average DOD acquisition program taking more than 2 years to become established and to reach an initial milestone, and as much as a decade to produce IOCs for testing, it is obvious that the pace associated with the DOD acquisition processes greatly lags that for IT advancement. It should also be noted that the nature of the current oversight process for DOD acquisition programs tends to encourage the aggregation of many requirements into larger programs, further exacerbating cycle time mismatches. This requirements aggregation is driven by such factors as a desire to avoid the cost of separate program documentation, a desire to minimize acquisition oversight reviews (see below), and a perceived need to consolidate budget requirements in order to raise the significance or importance of the program so that it will attract funding.

Another difference between weapons systems and IT systems is that of requirements specificity at various points in the development process. This difference manifests itself in two important ways: (1) user interaction and (2) boundary conditions.

With regard to user interaction, it does not make sense to develop detailed user-interaction requirements at the beginning of an IT program, particularly for a brand-new capability. IT systems provide capabilities that are very user-centric, and in some cases completely new user interaction must be observed and/or measured and program

[5] However, some IT programs both develop software and provide the hardware to operate it across a large production inventory. For these classes of programs, a significant part of the total cost is in the procurement of the hardware and its installation on a ship. This is one motivation for the discussion in Chapter 3 about breaking such programs up into their off-the-shelf-components (COTS hardware, software, and services [CHSS]) and the development and/or integration components (software development and commercial off-the-shelf integration [SCDI]) rather than having a single program.

flow or other system elements updated to reflect user preferences and usage models. Moreover, IT systems need to evolve as military missions change owing to threat or organizational changes as well as to improve the effectiveness and efficiency of military operations. Conversely, the evolution of weapons systems platforms reflects longstanding and time-honored operational methods that relate to the inherently long timescales involved in the design and building of platforms, in training, and so forth.

An early emphasis on setting detailed requirements often results in IT systems failing to meet user needs or meeting, too late, a user requirement that has long since changed. When this happens, military members often develop "homegrown" solutions in the field to meet their needs. Such a solution may address the user's immediate problem, but it creates new problems—such as even greater disconnects between centrally managed and centrally specified IT systems and user expectations, as well as a potential proliferation of poorly documented and supported solutions. One way to help avoid such a situation is to improve the coupling of the end user's perspective to development and testing, allowing the process to more formally embrace and harness edge innovation.

With regard to boundary conditions, IT systems have, at best, amorphous boundaries. For example, few if any operationally relevant IT systems components can operate independently of the network for long periods of time. Eventually each component must join the network to receive and relay information. IT systems capabilities tend to be widely dependent on (and often distributed across) the network. Moreover, these days joint and coalition operations are the rule rather than the exception, increasing the necessity of interconnection. When forces are interdependent, they need their IT systems to interoperate seamlessly in dynamic situations that cannot be easily forecast. From a user perspective, an IT component should be able to "do it all." As a result, requirements changes during and after development may be perceived by the user to be minor, but they may in fact compromise architectural decisions made very early in a program.

Conversely, weapons systems platforms tend to have physically discrete boundaries, defined very early on, that the user understands cannot be broached with impunity. In other words, the IT systems acquisition challenge is to fit large, fixed pieces together to construct, from a user perspective, a smoothly integrated whole, whereas many weapons systems platforms deal with much smaller chunks of commercial hardware and software and a great deal more customized hardware or developed software.

REQUIREMENTS PROCESS IMPEDES USE OF COMMERCIAL OFF-THE-SHELF SOLUTIONS

The use of COTS products has long been a staple in the commercial sector and is generally preferred over unique, one-of-a-kind systems development. COTS packages such as SAP Logistics, PeopleSoft for human resources, Siebel CRM for customer relationship management, and Oracle Applications for financial systems have successfully penetrated and provide day-to-day support to the commercial sector. Industry tends to adapt its business processes to standard COTS configuration templates so as to minimize time lines and costs associated with tailoring and long-term maintenance. Similarly, firms often find it advantageous to wait for features and functionality that exist on COTS roadmaps—or to engage with industry to get features added to those roadmaps—rather than to develop custom code or configurations. Conversely, the federal government has been slow to adopt COTS solutions, arguing that federal requirements are unique. Noncommercial requirements established by law and regulation that impose unique requirements exacerbate the problem. The resistance in the federal government to COTS solutions has softened over the past few years, but use of these kinds of COTS products in the federal government is still in the adoption phase.

The DOD has also been slow to adopt COTS products on the basis that "the DOD is different." When the DOD selects a COTS product for application, the acquisition process often devolves into a significant modification of the COTS product to meet DOD-unique requirements. A notable example of this contrast involves enterprise resource planning (ERP) systems. Although industry does not have a perfect record, the list of failed DOD ERP projects is very long. Vast sums of money have been spent on these failed attempts to modify or uniquely configure COTS IT products.[6]

Part of the impediment to the DOD's adoption of COTS solutions lies in its process for developing requirements. COTS products have processes, inputs, and outputs that are specifically defined. Industry has learned to adapt to these processes and products in the interests of economics and rapid development and deployment time lines. The DOD

---

[6] See the following reports from the Government Accountability Office, Washington, D.C.: GAO-08-927R, *DOD Systems Modernization: Maintaining Effective Communication Is Needed to Help Ensure the Army's Successful Deployment of the Defense Integrated Military Human Resources System*; GAO-08-822, *DOD Business Systems Modernization: Key Marine Corps System Acquisition Needs to Be Better Justified, Defined, and Managed*; GAO-08-896, *DOD Business Systems Modernization: Important Management Controls Being Implemented on Major Navy Program, But Improvements Needed in Key Areas*; and GAO-08-866, *DOD Business Transformation: Air Force's Current Approach Increases Risk That Asset Visibility Goals and Transformation Priorities Will Not Be Achieved.*

has, until recently, argued that the requirements of DOD systems cannot be easily adapted to COTS products, which in turn requires changes to the basic COTS software. This approach not only is counterproductive but also soon turns the ostensibly COTS solution into a one-of-a-kind system, weighed down with development, testing, deployment, and maintenance challenges. The result is that commodity IT technologies that underpin DOD IT systems emerge on 18-month cycles, but it takes several years for the average IT program to field an IOC (see Chapter 1).

For COTS acquisition, a governance model and associated processes focused on new development, such as the DOD 5000 series, are not a good fit. Specifically, many pre-Milestone B activities addressed in the DOD 5000 series do not apply to predominantly COTS-based DOD IT systems. For example, it does not make sense to require IT programs to conduct technology demonstrations of COTS components that are already in widespread use. The managers of both the DOD Network Centric Enterprise Services (NCES) and Network Enabled Command and Control (NECC) programs cited this as a problem in briefings to this committee.[7]

The alternative is to emphasize the effective purchasing of COTS products to meet enterprise standards (as opposed to program or project-unique needs) and the organizing of programs in an incremental, modular fashion. Governance then is focused on the horizontal integration of system services.

OVERLY LARGE INFORMATION TECHNOLOGY PROGRAMS INCREASE RISK

Current IT systems acquisition processes in the DOD encourage the bundling of many capabilities into a single program activity. A premise for such bundling is that once a team is in place, it will be able to deliver the desired capabilities without the overhead of assembling a new team. The aim is to reduce the learning curve and start-up times on both the DOD and the contractor sides. However, in the current acquisition process, aggregating what could have been several smaller development efforts into a larger major program means that the delivery time and costs are significantly increased and that the ability to leverage state-of-the-art technology is impeded. Previous studies (see Chapter 3) have shown that the evolving of a software solution in short-term, lightweight spirals, with user feedback from early fielded capabilities influencing successive spirals, is the most effective approach for IT systems. This approach is, of

[7] Timothy J. Harp, Deputy Assistant Secretary of Defense (C3ISR & IT Acquisition), "Information Technology Acquisition," presentation to the committee, Washington, D.C., February 25, 2009.

course, not without its own risks,[8] but it should be demanded and supported by the IT systems acquisition process. The size of projects has a significant impact on the ability to assess success and failure promptly. Program size can mask real operational problems. For example, IT standards compliance in large programs might have been accomplished only for limited portions of the overall supplied capability, yet the program in total is reported as compliant. Only during operational use will shortcomings that have developed be discovered. It is better to field smaller increments early and to discover such shortcomings when there is still the opportunity to fix them than to turn over an IT system for maintenance, only to discover severe shortfalls across the entire system.

FUNDING PROCESS IMPEDES FLEXIBILITY

The DOD's process for obtaining funding for new acquisition programs typically takes multiple years. To address a DOD capability shortfall, the shortfall must be linked to a request to Congress for funding that would be provided in a future year. For solutions that will rely on information technology, the time frame for seeking funding can be many times longer than the actual time needed to develop or procure the solution. If it is to achieve a more rapid delivery of information technology solutions, the DOD will need a more responsive process for justifying and allocating funding to address capability shortfalls.

Although government funding processes are controlled by Congress and have legally binding controls, there are a number of opportunities to use existing flexibility to achieve improved speed and agility. For example, there are authorities given to the DOD to allocate funds for urgent warfighter needs and to reallocate funding after congressional appropriation as a result of changing needs. These processes could be used to initiate IT solution development in weeks or months, leading to rapid fielding. In addition, in many cases acquisition funds are allocated by Congress to a larger mission or program area or in some cases to a portfolio of projects identified with an area of mission need—for example, to fund software upgrades in a particular mission area or system. The potential exists to leverage such flexibilities to establish an empowered and accountable process that could rapidly allocate funding to IT projects. Providing transparency to Congress regarding the use of these flexibilities would be

[8] One common scenario occurs when the early cycles deliver modest capability and the more challenging deliverables are deferred to later cycles. In such a case, either the program runs out of money and adequate capability is never delivered, or the difficulties of the system are not perceived until much too late. This is not a failure of the spiral approach per se but of governance and oversight. Nevertheless, a spiral approach does not guarantee success.

important in order to ensure that proper oversight is maintained and to demonstrate achievement of the objective of rapid delivery.

In the longer term, the DOD could work with Congress to establish a new set of funding mechanisms for IT-supported requirements that would align congressional funding with mission or capability areas rather than with individual acquisition programs. Under this concept, Congress would allocate funding to a mission area that would be governed in the DOD through a process similar to portfolio management. In implementing this concept, DOD officials would be responsible for setting priorities and allocating the funding to individual IT projects after the congressional appropriation of funds to a portfolio of mission requirements. This approach can ensure appropriate justification of funding needs tied to mission requirements during budget submission as well as the rapid allocation of appropriated funding consistent with the pace of evolving mission requirements and technology advancements. Currently the DOD uses a process similar to this concept for funding maintenance upgrades to aircraft avionics software. Likewise, a somewhat similar process is used for managing IT projects funded through working-capital funding processes. These and other examples of flexible and rapid funding processes should be useful models as the DOD works with Congress to establish a new funding process for acquisition of information technology.

EXCESSIVE OVERSIGHT, YET INSUFFICIENT PROGRAM ACCOUNTABILITY

It has often been observed that although the predictable response of an organization to program failures is to institute additional oversight, the burdens resulting from that oversight may paradoxically increase the likelihood of future failures. This phenomenon appears to have been at work in the DOD, where problems in past IT programs have led to oversight that delays program completion without necessarily ensuring the delivery of timely and useful capabilities.

In the current program environment, there can be multiple oversight bodies, and there are numerous participants in the program oversight and review process. Specifically, there are layers of integrated product teams (IPTs) that perform reviews of programs on a periodic basis and as a precursor to milestone decisions. These layers of IPTs consist of working-level IPTs (WIPTs), integrated IPTs (IIPTs), and overarching IPTs (OIPTs). When the IPT construct was originally instituted, it was intended to help a program resolve program issues quickly and at a level as low as possible in the organization. However, over time, the IPT structure has become a burdensome process often consisting of representation by organizations with special interests and with no accountability for program success.

Each of the oversight groups can require program changes, and often each individual representative of a Service or group has the ability to force changes to a project or to require special accommodations or requirements at a very detailed level, often without any justification with respect to impacts on the cost and/or schedule. In consequence, "process leadership" has replaced results-oriented IT acquisition leadership; success is defined as strict adherence to the acquisition process and to specific requirements defined at the outset of an acquisition program, even when the end-user capability is hopelessly compromised. Tracking the process milestones can easily become mistaken for the tracking of real project results. Real project progress, assessment, and milestone evaluation are often confused with adhering to process requirements. Ultimately, too many pocket vetoes result in the substitution of process for product.

Although processes are important, true operational measures of success for a program, especially those related to end-user satisfaction, are more important. Related to this, shared personal responsibility and accountability on the part of all participants in the process are essential for program success. This is especially important in joint programs, where Service-specific preferences have to be reconciled with the common good.

There are notable counterexamples to the trends discussed above. Several large and complex DOD programs without high degrees of institutional oversight have been demonstrably successful in rapidly delivering useful capability to the field by following tailored, focused, proactive, accountable oversight of the kind advocated in this report; see the examples given in Box 2.1 and detailed in Appendix D. Such programs tend to have been managed by program managers (PMs) who figured out how to navigate the acquisition process effectively so that they could focus on meeting end-user needs and keep oversight to a reasonable level. The key is in striking the right balance—placing on the program manager appropriate responsibility and accountability for meeting end-user needs in a timely way while establishing appropriate levels of oversight.

One way to address this challenge is to explicitly clarify and strengthen the responsibility, authority, and accountability for program execution and to create a process that allows for clear, more timely, and more accurate assessment of a project's progress and risk and for the early identification of failing projects. One approach would be to provide the PM and a portfolio management team (PMT) with decision authority, derived explicitly from higher authority, to determine trade-offs among schedule, cost, and functionality. The PMT would represent several equities—an acquisition equity at or above the next echelon up from the PM, a functional equity representing the voice of the end user, and an enterprise equity typically represented by the chief information officer or the CIO's designee. The

BOX 2.1
Succeeding with Nontraditional Oversight

Selective examples of large and complex Department of Defense information technology (IT)-based programs that were demonstrably successful with nontraditional oversight are identified below and discussed in detail in Appendix D. Collectively, these success stories provide evidence that changes of the nature proposed in this report can have dramatically positive impacts. Common characteristics among these programs include the following:

- Support by a senior "champion" or advocate providing "top cover";
- Urgent operational need;
- Their iterative incremental development approach;
- The early and continuing involvement of the end users in the development process;
- Field experiments with early testing, rapid feedback, and rapid fixes (e.g., Advanced Concept Technology Demonstration of Army Digitization Experiments);
- Initial management at a lower-level acquisition category (including those programs started as non-program of record [POR] initiatives but subsequently formalized as a POR after initial operating capabilities [IOCs] were deployed);
- The use of "agile" approaches to deliver capability via software increments deployed on an existing infrastructure;
- Substantial leveraging of commercial off-the-shelf (COTS) solutions with waivers or work-arounds to meet military-specific requirements or desires;
- Contractor logistics support at least through IOC deployment;
- Leveraged supplemental funding or "modification-in-service" funding lines that had been established to provide predictable annual resources for upgrades and technology insertion in fielded IT-based systems. Such programs may have started with oversight at the Office of the Secretary of Defense level, but sometime after IOC, oversight authority was delegated to the Services or agencies.

Following are some notable examples:

- *Force XXI Battle Command Brigade and Below (FBCB2).* This command-and-control system was developed as a central element of the Army's Advanced

PMT, in consultation with the PM, would be empowered to make decisions about such things as development priorities, the contents of capability increments, and what constitutes "must have" versus "nice to have."

The examples in Box 2.1 also suggest several other ways to achieve greater agility and responsiveness and to avoid the stifling oversight

Warfighting Experiments (AWE) in the mid-1990s. A companion development was the Army's Tactical Internet, a data communications capability designed by adding commercial router technology to legacy tactical communications devices. This capability was developed with a user jury of warfighters in the loop. It has been extraordinarily successful on the battlefield. More than 40,000 systems are fielded today.

- *Blue Force Tracker (BFT).* This program is a variant of FBCB2 that uses satellite-based communications in lieu of terrestrial communications capabilities. Early variants were deployed on surrogate commercial computers for use during the conflict in the Balkans in the late 1990s. During 2002 an intensive effort was initiated with supplemental funds to develop and deploy BFT for forces being prepared for Operation Iraqi Freedom.
- *Joint Network Node (JNN).* This program was accelerated through the use of a modification-in-service funding line, in this case the funding line for the Army's Mobile Subscriber Equipment program. Operating with decentralized oversight, the program rapidly delivered COTS- and government-off-the-shelf-based solutions for forces deploying to Iraq. The program was so successful that JNN systems were fielded throughout the Army. Eventually it was elevated to an acquisition category (ACAT) 1D program because its cost reached levels warranting that designation.
- *Command Post of the Future (CPOF).* This is a command-and-control program built with advanced visualization and collaboration technology from the commercial and academic sectors that was initiated by the Defense Advanced Research Projects Agency (DARPA). There was early collaboration with the user community during system development, and the system was deployed for evaluation, training, and interoperability enhancements at the Central Technical Support Facility at Fort Hood, Texas. Based on its success, the system was transitioned into a formal POR for support, further fielding, and upgrades to meet evolving end-user requirements.
- *Tactical Ground Reporting System (TIGR).* This is another DARPA program. It is a multimedia reporting system for soldiers at the patrol level, allowing users to collect and share information to improve situational awareness and to facilitate collaboration and information analysis among junior officers. It is based on commercial information technology and was developed using rapid and agile acquisition processes without going through the normal oversight process. It was developed in collaboration with end users and has evolved into a highly valued, widely deployed system in Iraq and Afghanistan.

inherent in the preponderance of acquisition category (ACAT) I-level programs (see the section below on "Legislative Impediments" for information on ACAT programs). One of the best channels for doing so is to leverage the opportunity to insert improvements on top of existing platforms. The conundrum is how to get a baseline initial operating capability in place so that it can be used as a platform for rapid development and fielding. Large programs must be structured to move to IOC in a more timely manner so that they can become the "platforms" on which future capabilities can be acquired in an agile, iterative, responsive fashion to meet end users' ever-changing requirements and to take advantage of emerging technologies.

The biggest challenges in effecting real change will continue to be with "new starts" and larger-value DOD IT programs for which there are demands for detailed justification, a ponderous programming and budgeting process to get multiyear funding justified in the Program Objectives Memorandum and appropriated annually, and "disciplined" ACAT I-level oversight processes that do not align well with agile acquisition.

Another way to foster greater agility is to structure the DOD IT portfolio so that it includes a small number of true enterprise-level programs that are so big, so important, and so costly (i.e., above the funding thresholds at which weapons systems are categorized as ACAT ID) that they warrant intensive management and oversight at the Office of the Secretary of Defense (OSD). A good example of a large enterprise-level IT program is the Global Information Grid-Bandwidth Expansion program. Such large programs could be scheduled so that they deliver next-generation platforms at a point in time when the legacy platforms are nearing the stage at which they can no longer be sustained through annual upgrades and should be replaced. If the remaining programs in the DOD IT portfolio were structured for decentralized management and oversight, the IT system acquisition community would be better positioned for agile acquistion. Decentralized programs might be categorized in the DOD IT portfolio as (1) modification-in-service programs and (2) new programs structured at dollar levels that permit designation as ACAT II and ACAT III with the same dollar thresholds used for weapons systems and with oversight by the Services, agencies, or program executive officers.

CULTURAL IMPEDIMENTS TAKE PRECEDENCE OVER RAPID DEVELOPMENT

The DOD's perceived need for caution over speed is understandable. Given the criticality and danger of its mission, its worldwide operations and large workforce, and the frequent need for clear, decisive action, the Department of Defense, by its nature, is an organization with a classic command-and-control culture. If current trends continue, it is likely that processes and systems will become even more top-down and centralized in spite of the DOD's desire to move to an integrated, cross-Service environment with empowered decision making at all levels of command. Although current doctrine is shifting from a "need to know" basis to a "need to share" basis, it is being accomplished through clearly controlled and hierarchical processes and systems.

In a command-and-control environment, the steps in an IT system's life-cycle development process are based on frequent reviews and concurrence by a large number of concerned, but often narrowly focused, stakeholders. Such an environment does not lend itself to rapid innovation or to rapid development processes.

Meaningful assessment becomes nearly impossible when large, complex programs have long time spans between significant milestones. Current DOD processes, for example, put great emphasis on detailing requirements before a program is approved to start, so that costs and risks can be evaluated. Although this makes sense for most weapons systems, for IT systems it often results in years of requirements development, leading to the delivery of IT systems that are trying to meet requirements that have long since changed or are continuing to shift. A great deal of project time can pass while costs accumulate before a meaningful assessment of project viability can be made.

To fundamentally recalibrate the culture, any proposed new DOD IT systems acquisition process should focus on the rapid fielding of successive increments of capability. The time of delivery of some useful and usable capability should become a key performance parameter. A rapid delivery capability based on commercially available technologies is needed, along with a fielding model that permits the evolution of these capabilities as the technology evolves during training, integration, maintenance, and support. The DOD must move to a culture that enables the rapid adoption of new technologies and a process that assesses where such rapid developments are essential. Delivering increasingly useful increments of capability should receive a higher evaluation than that given for delivering nothing to the field for several years.

In conjunction with this process, necessary to program success is a culture of "tough love" communicated through direct, pragmatic advice that is given by oversight organizations to the program managers in combination with the element of advocacy. Such a culture needs to replace excessive oversight, a "gotcha" mentality, and gatekeeping. The culture must support the whole team in the effort to help the PM deliver capabilities to the user, replacing a culture focused on rigid controls, fault finding, and grading of the PM's work.

More generally, incentivizing the use of IID and agile-inspired approaches to acquisition and development is key. The use of successful IID approaches and the delivery of a capability to the end user should merit rewards for the whole team. It will be important to foster the notion that when a program "wins," the whole team wins. (See Chapter 3 for a discussion of how the handling of failure also needs to change.)

Leadership must come from senior levels within IT organizations, not only from the IT staff. Defense acquisition executives have an important role to play in establishing a culture that encourages IID. In addition, a process by which successes and failures are reviewed on a relatively frequent periodic basis would aid in helping teams understand what is working and why (or why not). In essence, an iterative approach can be applied to the process as well as to the technology.

INADEQUATE INFORMATION TECHNOLOGY ACQUISITION WORKFORCE

Under the leadership of the Under Secretary of Defense for Acquisition, Technology and Logistics, and as specified in DOD acquisition regulations, the DOD Acquisition Corps is charged with procuring systems and services to meet warfighters' needs in a timely fashion as required to satisfy national security objectives. The Acquisition Corps includes many highly trained specialists in areas of engineering, science, testing, and business and program management who act as acquisition executives, program managers, and contracting officers. A number of studies have expressed concern about the technical proficiency of the acquisition workforce.[9] Over the past two decades, numerous defense authorization and appropriation bills have included provisions aimed at improving the training of acquisition professionals.

Among the many challenges in this area is that relatively few in the acquisition workforce have specific expertise in IT or in how to manage IT programs. An important factor is that there are few "digital natives"—people who have grown up with and/or are highly proficient with IT—in the ranks of senior acquisition PMs.

Another factor is that very little currently available formal training is focused on the distinct issues that arise in DOD IT programs. Although the Defense Acquisition University (DAU), the premier source of acquisition training in the DOD, offers both resident and remote training programs that emphasize systems program management and policy compliance, the DAU does not have a comprehensive program to teach IT program management or IT test and evaluation.

As a result, the acquisition workforce is not well equipped to manage IT programs. Because the DOD's acquisition regulations were designed primarily to meet the needs of large weapons programs, significant tai-

---

[9] Several reports on this topic are reviewed in Government Accountability Office, *Contract Management: DOD Vulnerabilities to Contracting Fraud, Waste, and Abuse*, GAO-06-838R, Washington, D.C., July 2006. That report also summarizes other challenges facing the acquisition system.

loring is required to accommodate IT programs, especially to follow the incremental, iterative development process recommended in this report. Without a nuanced understanding of IT and the needs of IT programs, an IT program manager is thus at a disadvantage in advising or embarking on the tailoring of acquisition processes. Even those PMs willing and able to advocate for significant tailoring would be incurring additional risk by embarking on a course distinct from the standard acquisition process.[10]

Moreover, personnel practices that are common in the acquisition community make it nearly impossible to align rewards and penalties with true program success. Contributing factors include these:

- The DOD rotates personnel too often for any one PM to see an acquisition through more than a single milestone;
- The acquisition process rewards the following of acquisition processes rather than the delivery of useful and usable capability to end users;
- The military culture is a "can do" culture—no program manager wants to say that a given task cannot be done; and
- Program size is used as a success metric and is associated, overtly, with rank. As a result, program managers are incentivized to make programs larger, which contrasts starkly with evidence from many studies that smaller programs reduce cost and risk.

Closely related to acquisition workforce capabilities is the DOD's capacity to be a smart buyer—to possess the in-house technical expertise (that is, scientific, engineering, and mathematical skills) required to engage effectively with industry on technical design, research and development, and procurement matters. It is generally viewed as inherently a government responsibility that cannot be delegated to industry. As such, sustaining (and enhancing) a smart-buyer capability is a necessary complement to efforts to strengthen the acquisition workforce or to reform DOD acquisition processes.[11]

---

[10] For example, a PM fielding a COTS-based capability might attempt to argue that the Technical Readiness Assessment pertinent to the program should assess the integrator's ability to build and deploy components based on past performance rather than the COTS ability to support common infrastructure requirements. In today's environment, the likely result of such a discussion would require the PM to do both: demonstrate that widely deployed COTS works as advertised *and* provide data supporting the integrator's ability to build and deploy components on the infrastructure.

[11] See, for example, Kenneth Horn, Carolyn Wong, Elliot Axelband, Paul Steinberg, and Ike Chang, *Maintaining the Army's "Smart Buyer" Capability in a Period of Downsizing*, RAND, Santa Monica, Calif., 1999. Available at http://www.rand.org/pubs/white_papers/2005/WP120.pdf; accessed December 12, 2009.

LEGISLATIVE IMPEDIMENTS

Today's acquisition oversight process in the DOD is designed for the disciplined management of large, expensive, complex weapons systems—a process whose overall features are dictated by statute. Programs are assigned acquisition categories based on acquisition cost estimates and are designated for oversight levels based on associated cost thresholds—at the Office of the Secretary of Defense, at the Service or agency level, or at lower levels such as that of program executive officers. However, the total dollar thresholds for designating oversight levels for IT programs are significantly lower than those used for weapons systems (by a factor of five).[12] This results in a dichotomy in which an IT system with a development and deployment cost of $126 million over its life cycle has highly centralized oversight at OSD, while a weapons system counterpart at the same dollar level can be decentralized for oversight at the program executive officer level. Moreover, the current legislation has no provision for major automated information system (MAIS) programs to receive oversight at the Service or agency level. One approach to solving the problem of highly centralized oversight with its attendant delays would be to use the same dollar thresholds in effect for major defense acquisition programs (MDAPs) for the designation of ACAT levels to MAIS programs. This elevation of thresholds for IT programs would better align the authority for IT program oversight to the appropriate levels at OSD, the Services and agencies, and lower echelons.

MEASURES OF SUCCESS

The committee fully anticipates that any new IT systems acquisition process that is defined and adopted will, by definition, evolve over time. The committee has identified shortcomings in the present system and recommends a new direction for and approach to IT acquisition. Evaluating the progress and success of programs is a critical component of this approach. Indeed, for each program there should be a tailored set of metrics that are agreed to.

This report does not recommend a particular set of metrics; instead it describes the generic categories from which the metrics should be drawn. For example, there will be instances in which an IID approach is aimed at

[12] ACAT I programs are those estimated to require eventual expenditure for research, development, test and evaluation of more than $365 million or procurement for more than $2.19 billion. Major automated information system (MAIS) programs are those estimated to require program costs in any single year in excess of $32 million, total program costs in excess of $126 million, or total life-cycle costs in excess of $378 million (all figures in FY 2000 constant dollars).

developing modules that will reside within a well-established framework that has already been developed to have security, reliability, and/or other nonfunctional characteristics, and elaborate metrics will not be required in those areas. But there will be other cases where the infrastructure is not available or is inadequate, and in those cases the metrics will need to be expanded to account for the infrastructure requirements as well as the intended functionality. Similarly, although the dominant focus should be on end-user needs, attention should also be paid to other stakeholders as appropriate. The following measures of success for IT systems acquisition are suggested as a useful set of guidelines to consider when developing tailored metrics for particular programs:

- End-User Capability
  —Measurable improvements in currently fielded end-user capability, including functionality, performance, the meeting of commercial benchmarks, and reliability experienced by the end user.
  —Measurable reduction in the costs of currently fielded IT operations.
- End-User Satisfaction
  —Measurable improvement in end-user satisfaction.
  —Measurable increase in consumption (use by end users).
- Timeliness
  —Significantly reduced COTS fielding times.
  —For software development and commercial off-the-shelf integration programs, fielding capability within the product cycle of major COTS components.
  —For COTS hardware, software, and service programs, fielding pace comparable to COTS cycles (for example, no more than 12 to 18 months for COTS hardware).
  —Significantly reduced increment cycle times—no more than 12 to 18 months between increments of fieldable end-user capability.
- Quality
  —Measurable decreases in the number and severity of bugs discovered postfielding (including security vulnerabilities).
  —Measurable improvements in availability and reliability as experienced by the end user.
- Operating Costs
  —Measurable reduction in currently fielded IT operations costs.
  —Measurable improvements in administrator-to-server ratios.
  —Measurable improvements in server-to-client ratios along with a decrease in the unit costs of bytes served (output delivered).

- Acquisition
  —Measurable improvements in progress against budget, schedule, and functional capability.
  —Demonstrably part of a long-term competitive strategy.

Corresponding, specific target metrics could be developed for each portfolio and project or program funded by the portfolio at the initiation of a program and defined incrementally for each iteration within a project or program. (For example, a performance metric might be tightened over successive increments of a program.) These metrics would form the basis for reporting progress to senior DOD and Office of Management and Budget officials as well as to Congress.

3

Systems and Software Engineering in Defense Information Technology Acquisition Programs

THE EVOLUTION OF DEPARTMENT OF DEFENSE POLICY AND PRACTICE FOR SOFTWARE DEVELOPMENT

Beyond the sometimes burdensome nature of the oversight process as described earlier in this report and the inordinate amount of time that it can take an information technology (IT) program to reach initial operating capability (IOC), there are other fundamental issues afflicting many Department of Defense (DOD) IT programs. Many programs fail to meet end-user expectations even after they do finally achieve IOC. This, of course, is not a phenomenon unique to DOD IT programs, but it is certainly exacerbated by the long cycle times associated with DOD acquisition, especially for programs that exceed the dollar threshold for which department-level oversight is required. Such programs are designated as a major automated information system (MAIS) programs in the case of IT and as major defense acquisition programs (MDAPs) for weapons systems.

In the DOD, a significant structural factor leading to this failure to meet end-user expectations is the persistent influence over many decades of what is characterized as the waterfall software development life cycle (SDLC) model—despite a body of work that is critical of the waterfall mentality, such as the Defense Science Board reports cited below, and the issuance of directives identifying models other than the waterfall approach as the preferred approach. The waterfall model discussed below remains at least implicit in the oversight structure and processes that govern IT acquisition programs in the DOD today.

The waterfall process model for software development has its origins in work by Winston Royce in 1970.[1] The term *waterfall* refers to a sequential software development process in which a project flows downward through a series of phases—conception, initiation, analysis, design, construction, testing, and maintenance. Ironically, Royce was not advocating the waterfall model in his original paper even though the model is attributed to him. He cited the waterfall model as a commonly used but flawed development approach and instead advocated a "do-it-twice" iterative process with formal customer involvement at multiple points in the process as a means to mitigate risk in large software development projects.

A paper by Reed Sorenson outlines the evolution of DOD SDLC models in the subsequent decades.[2] The early years of that evolution were dominated by military standards such as Military Standard 490 on specification practices and DOD Standard (STD) 1679A on software development. Although DOD-STD-1679A was focused on software development, its origins in hardware and weapons systems development are clearly evident, and it reflects an era in which the waterfall SDLC model predominated.

The evolution continued through DOD-STD-2167 and 2167A in the late 1980s, driven in part by strong criticism of the waterfall model and a growing appreciation for a model not so heavily influenced by hardware and weapons systems thinking. Brooks's seminal paper "No Silver Bullet—Essence and Accidents of Software Engineering," published in 1987, was among the first to criticize the notion integral to the waterfall model—specifically, that one can fully specify software systems in advance:

> Much of present-day software-acquisition procedure rests upon the assumption that one can specify a satisfactory system in advance, get bids for its construction, have it built, and install it. I think this assumption is fundamentally wrong, and that many software-acquisition problems spring from that fallacy.[3]

A 1987 Defense Science Board study chaired by Brooks was equally critical of the waterfall mentality contained in the then-in-force DOD-STD-2167 and the then-draft DOD-STD-2167A:

---

[1] Winston Royce, "Managing the Development of Large Software Systems," pp. 1-9 in *Proceedings of the IEEE Westcon*, IEEE, Washington, D.C., 1970.

[2] Reed Sorenson, "Software Standards: Their Evolution and Current State," *Crosstalk* 12:21-25, December 1999. Available at http://www.stsc.hill.af.mil/crosstalk/frames.asp?uri=1999/12/sorensen.asp; accessed December 12, 2009.

[3] F.P. Brooks, Jr., "No Silver Bullet—Essence and Accidents of Software Engineering," *Information Processing* 20(4):10-19, April 1987.

> DOD Directive 5000.29 and STD 2167 codify the best 1975 thinking about software including a so-called "waterfall" model calling for formal specification, then request for bids, then contracting, delivery, installation and maintenance. In the decade since the waterfall model was developed, our discipline has come to recognize that setting the requirements is the most difficult and crucial part of the software development process, and one that requires iteration between designers and users.[4]

In 1985 Barry Boehm first published his "spiral model" for software development,[5] driven in part by this same fundamental issue: the ineffectiveness of the traditional waterfall software development process model. The Defense Science Board, in reports in 1994[6] and 2000,[7] continued to argue for the abandonment of the waterfall model, the adoption of the spiral model, and the use of iterative development with frequent end-user involvement.

In 2000, DOD Instruction (DODI) 5000.2 was revised. For the first time the acquisition policy directives identified evolutionary acquisition as the preferred approach for acquisition. In 2002 the Under Secretary of Defense for Acquisition, Technology and Logistics issued a memorandum clarifying the policy on evolutionary acquisition and spiral development and setting forth a model based on multiple delivered increments and multiple spiral cycles within each delivered increment.

The current version of DODI 5000 retains the policy statement that evolutionary acquisition is the preferred approach. It further provides the governance and oversight model for evolutionary development cycles. However, the 5000 series regulations remain dominated by a hardware and weapons systems mentality. For example, the terminology used to describe the engineering and manufacturing development phase emphasizes the hardware and manufacturing focus of the process. In the evolutionary acquisition governance model, each phase repeats every one of the decision milestones A, B, and C and also repeats every program phase.

---

[4] F.P. Brooks, Jr., V. Basili, B. Boehm, E. Bond, N. Eastman, D.L. Evans, A.K. Jones, M. Shaw, and C.A. Zraket, *Report of the Defense Science Board Task Force on Military Software*, Department of Defense, Washington, D.C., September 1987.

[5] Barry Boehm, "A Spiral Model of Software Development and Enhancement," *Proceedings of the International Workshop on Software Processes and Software Environments*, ACM Press, 1985; also in *ACM Software Engineering Notes* 15(5):22-42, August 1986; and *IEEE Computer* 21(5):61-72, May 1988.

[6] Department of Defense, *Report of the Defense Science Board on Acquiring Defense Software Commercially*, June 1994; available at http://www.acq.osd.mil/dsb/reports/commercial defensesoftware.pdf; accessed December 12, 2009.

[7] Department of Defense, *Report of the Defense Science Board Task Force on Defense Software*, November 2000; available at http://www.acq.osd.mil/dsb/reports/defensesoftware.pdf; accessed December 12, 2009.

Preliminary design reviews (PDRs) and critical design reviews (CDRs), hallmarks of the waterfall SDLC model, are prescribed for every program, with additional formal Milestone Decision Authority (MDA) decision points after each design review. At least four and potentially five formal MDA reviews and decision points occur in every evolutionary cycle. As a result, although the oversight and governance process of DODI 5000 does not forbid the iterative incremental software development model with frequent end-user interaction, it requires heroics on the part of program managers (PMs) and MDAs to apply iterative, incremental development (IID) successfully within the DODI 5000 framework. (A separate question not addressed by this report is whether the current process is well suited for weapons systems.) Moreover, the IID and evolutionary acquisition approaches address different risks (Box 3.1).

Today, many of the DOD's large IT programs therefore continue to adopt program structures and software development models closely

BOX 3.1
Evolutionary Acquisition Versus Iterative, Incremental Development

Because both are incremental development approaches, evolutionary acquisition (EA) and iterative, incremental development (IID) are sometimes confused, even though the motivation for each approach and the nature of the increments used in each approach are different. The EA approach is motivated by a need for technology maturation—early increments provide end-user capabilities based on mature technology, whereas work on later increments is deferred until needed technology has been matured. In contrast, IID development for information technology (IT) systems is based on mature technology—that is, no science and technology development is needed. For many IT systems, important user-interaction requirements cannot be defined in detail up-front and need to be determined and verified based on incremental feedback from users. Accurate feedback cannot be provided unless users interact with actual systems capabilities. Increments in IID for IT systems are intended to provide the basis for this requirements refinement; experience shows that without such testing in real-world environments, a delivered IT system is unlikely to be useful to its intended users without the undertaking of extensive reworking.

The EA approach is particularly useful when the technology required to support a needed capability is not completely mature—those capabilities that can be provided on the basis of mature technology are implemented in initial increments; the technology needed for later increments is matured while work on the initial increments proceeds.

An important difference in the two approaches is the timescale for the increments. For IT systems, each increment should deliver usable capability in less than 18 months. DODI 5000.2 suggests that an increment should be produced in fewer than 5 years (for weapons systems).

resembling the waterfall model rather than an IID model with frequent end-user interaction. Even those that plan multiple delivered increments typically attempt to compress a waterfall-like model within each increment.

Largely the same governance and oversight approach is applied to IT programs as that applied to weapons systems programs except for the financial thresholds used to designate an IT program as one meriting department-level oversight (in DOD parlance, a major automated information system). The expenditure levels necessary for an IT system program to be designated as an MAIS and subjected to very "heavyweight" governance and oversight processes are significantly lower than the expenditure levels necessary for a weapons system program to be designated as an MDAP and subjected to this same level of scrutiny.

This heavyweight governance process includes large numbers of stakeholders whose single-issue equities must be satisfied at each decision point or assessment. As discussed in Chapter 1 of this report, this governance and oversight process results in very long time lines that are fundamentally out of alignment with the pace of change in information technology. And although many stakeholders are permitted to express opinions at each assessment or milestone decision point in the governance and oversight process, the voice of the end user is seldom heard in this process.

ITERATIVE, INCREMENTAL DEVELOPMENT

This section begins with a brief history of iterative, incremental development drawn from Craig Larman and Victor Basili and from Alan MacCormack.[8] It is provided as a way to situate agile and related approaches within a broader context and also to demonstrate that IID has a long history of being applied successfully for different types and scales of problems. The section continues with a discussion of various agile software methodologies and their core concepts for how to better and more efficiently develop and acquire IT systems that meet end-user needs.

Emerging from a proposal by Walter Shewhart at Bell Labs, IID began as a quality-control exercise, encapsulated in the phrase plan-do-study-act (PDSA). PDSA was heavily promoted in the 1940s by W. Edwards Deming and was explored for software development by Tom Gilb, who developed the evolutionary project management (Evo) process in the 1960s, and by Richard Zulmer. The first project to actively use an IID process success-

[8] Craig Larman and Victor R. Basili, "Iterative and Increment Development: A Brief History," *IEEE Computer*, pp. 47-56, June 2003; and Alan MacCormack, "Product-Development Practices That Work," *MIT Sloan Management Review* 42(2):75-84, 2001.

fully was the 1950s X-15 hypersonic jet. IID was later used to develop software for NASA's Project Mercury. Project Mercury was developed through time-boxed,[9] half-day iterations, with planning and written tests before each micro-increment.

In 1970, Royce's article on developing large software systems, which formalized the strict-sequenced waterfall model, was published.[10] Royce actually recommended that the phases which he articulated—analysis, design, and development—be done twice. In addition to repetition (or iteration), Royce also suggested that projects with longer development time frames first be introduced with a shorter pilot model to examine distinctive and unknown factors. Although not classically IID, Royce's proposed model was not as limited as the waterfall model would become.

Drawing on ideas introduced in Project Mercury, the IBM Federal Systems Division (FSD) implemented IID in its 1970s-era software projects and enjoyed success in developing large, critical systems for the Department of Defense. The command-and-control system for the U.S. Trident submarine, with more than 1 million lines of code, was the first highly visible application with a documented IID process. The project manager, Don O'Neill, was later awarded the IBM Outstanding Contribution Award for his work in IID (called integration engineering by IBM). IBM continued its success with IID in the mid-1970s, developing the Light Airborne Multi-Purpose System (LAMPS), part of the Navy's helicopter-to-ship weapons system. LAMPS is notable for the fact that it used approximately 1-month iterations, similar to current popular IID methods of 1 to 6 weeks. LAMPS was ultimately delivered in 45 iterations, on time and under budget.

IBM's FSD also had success in developing the primary avionics systems for NASA's shuttle. Owing to shifting needs during the software development process, engineers had to abandon the waterfall model for an IID approach, using a series of 17 iterations over 31 months.

TRW, Inc., also used the IID approach for government contracts in the 1970s. One noteworthy project was the $100 million Army Site Defense software project for ballistic-missile defense. With five longer cycles, the project was not specifically time-boxed and had significant up-front specification work; however, the project was shifted at each iteration to respond to customer feedback. TRW was also home to Barry Boehm, who in the 1980s developed the spiral model described earlier in this chapter.

The System Development Corporation also adopted IID practices

---

[9] *Time-boxing* refers to a deadline-driven approach to systems development whereby work items may slip from one iteration to the next, but iterations are completed according to schedule, thus affording the opportunity to identify erroneous estimates of work items quickly and ensuring continuous user input regarding priorities.

[10] Winston Royce, "Managing the Development of Large Software Systems," pp. 1-9 in *Proceedings of the IEEE Westcon*, IEEE, Washington, D.C., 1970.

while building an air defense system. Originally expected to fit into the DOD's waterfall standard, the project was built with significant up-front specifications followed by incremental builds. In a paper published in 1984, Carolyn Wong criticized the waterfall process: "Software development is a complex, continuous, iterative, and repetitive process. The [waterfall model] does not reflect this complexity."[11]

The early1980s led to numerous publications extolling the virtues of IID and criticizing the waterfall approach. In 1982, the $100 million military command-and-control project, based on IBM's Customer Information Control System technology, was built using an IID approach without the time-boxed iterations, called evolutionary prototyping. Evolutionary prototyping was used often in the 1980s in active research in artificial intelligence systems, expert systems, and Lisp machines. During the mid-1980s, Tom Gilb was active in supporting a more stringent IID approach that recommended short delivery times, and in 1985 Barry Boehm's text describing the spiral model was published. In 1987 TRW began building the Command Center Processing and Display System Replacement. Using an IID system that would later become the Rational Unified Process, the 4-year project involved six time-boxed iterations averaging 6 months each. In the 1990s, the developers shifted away from the heavy up-front speculation still being used in the IID approach of the 1970s and 1980s. In 1995 at the International Conference on Software Engineering, Frederick Brooks delivered a keynote address titled "The Waterfall Model Is Wrong!" to an audience that included many working on defense software projects.

After near failure using a waterfall method, the next-generation Canadian Automated Air Traffic Control System was successfully built based on a risk-driven IID process. Similarly, a large logistics system in Singapore was faltering under the waterfall process. Jeff de Luca revived the project under IID and created the Feature Drive Development (FDD) process. According to a 1998 report from the Standish Group, the use of the waterfall approach was a top reason for project failure in the 23,000 projects that the group reviewed.[12]

Commercially, Easel Corporation began developing under an IID process that would become Scrum, an agile software development framework. Easel used 30-day time-boxed iteration based on an approach used for nonsoftware products at major Japanese corporations. Institutionalizing its own IID process, Microsoft Corporation introduced 1-day iterations. Beginning in 1995, Microsoft, driven to garner a greater share of the

---

[11] Carolyn Wong, "A Successful Software Development," *IEEE Transactions on Software Engineering* 3:714-727, 1984.

[12] The Standish Group, *Chaos: A Recipe for Success*, Boston, Mass., 1998.

browser market from Netscape Communications Corporation, developed Internet Explorer using an iterative, component-driven process. After integrating component modules into a working system with only 30 percent functionality, Microsoft determined that it could get initial feedback from development partners. After the alpha version of Internet Explorer was released, code changes were integrated daily into a complete product. Feedback was garnered in less than 3 hours, and so adjustment and new functionality could be added. With only 50 to 70 percent functionality, a beta version was released to the public, allowing customers to influence design while developers still could make changes.

Agile software development (ASD) methodologies, a form of IID, have been gaining acceptance among mainstream software developers since the late 1990s. ASD encourages the rapid development of systems with a high degree of requirements volatility. In addition, ASD practices emphasize delivering incremental working functionality in short release cycles. "The Manifesto for Agile Software Development" (Box 3.2) presents one group's view of ASD's strengths. Figure 3.1 shows Boehm's view of how ASD fits in the spectrum of development processes. The processes used by the DOD for IT acquisition traditionally fall in the right-hand side of the figure.

Today, ASD methodology is established to varying degrees in the academic, educational, and professional software development communities. Particular instantiations include the Scrum, Crystal, and Extreme Programming (XP) approaches. In 2001 the first text on the subject was published under the title *Agile Software Development*, by Alistair Cockburn.[13]

Several commercial IT companies are moving toward developing software using ASD processes. For example, a recent randomized survey of 10 percent of the engineers at Microsoft found that around one-third of the respondents use ASD. That survey also found that Scrum is the most popular ASD methodology, that ASD is a relatively new phenomenon to Microsoft, that most projects have employed ASD techniques for fewer than 2 years, and that ASD is used mostly by co-located teams who work on the same floor of the same building.

A fair amount of research has been conducted on teams adopting ASD processes.[14,15] From the academic research perspective, Laurie Williams

---

[13] A. Cockburn, *Agile Software Development*, Addison-Wesley, Boston, Mass., 2001.

[14] See http://www.controlchaos.com/resources/ for Scrum case studies and several examples of waterfall teams moving to Scrum; accessed November 2, 2009.

[15] See http://agile2009.agilealliance.org/; accessed November 2, 2009. The Agile conference series has various experience reports on teams adopting Agile. Agile 2006 and Agile 2007 sold out, with more than 1100 (predominantly industrial) attendees.

BOX 3.2
Manifesto for Agile Software Development

We are uncovering better ways of developing
software by doing it and helping others do it.
Through this work we have come to value:

Individuals and interactions over processes and tools
Working software over comprehensive documentation
Customer collaboration over contract negotiation
Responding to change over following a plan

That is, while there is value in the items on
the right, we value the items on the left more.

Kent Beck
Mike Beedle
Arie van Bennekum
Alistair Cockburn
Ward Cunningham
Martin Fowler
James Grenning
Jim Highsmith
Andrew Hunt
Ron Jeffries
Jon Kern
Brian Marick
Robert C. Martin
Steve Mellor
Ken Schwaber
Jeff Sutherland
Dave Thomas

NOTE: The Web site on which this manifeso appears contains the following notice: "© 2001, the above authors. This declaration may be freely copied in any form, but only in its entirety through this notice."
SOURCE: Available at http://www.agilemanifesto.org/.

and her research group[16] have conducted empirical evaluation of companies adopting XP. This work is summarized in Table 3.1, which highlights the research results of four case studies performed on small to medium-size systems. Across all systems, one sees a uniform improvement in post-

[16] See Laurie Williams, William Krebs, Lucas Layman, and Annie I. Antón, "Toward a Framework for Evaluating Extreme Programming," *Proceedings of the 8th International Conference on Empirical Assessment in Software Engineering,* Edinburgh, Scotland, May 24-25, 2004, pp. 11-20; Lucas Layman, Laurie Williams, and Lynn Cunningham, "Motivations and Measurements in an Agile Case Study," *Journal of System Architecture* 52(11):654-667, 2006; Lucas Layman, Laurie Williams, and Lynn Cunningham, "Exploring Extreme Programming in Context: An Industrial Case Study," *Proceedings of the 2nd Agile Development Conference,* Salt Lake City, Utah, pp. 32-41, June 2004; and Lucas Layman, Laurie Williams, and Lynn Cunningham, "Exploring Extreme Programming in Context: An Industrial Case Study," *Proceedings of the 2nd Agile Development Conference,* Salt Lake City, Utah, pp. 32-41, June 2004.

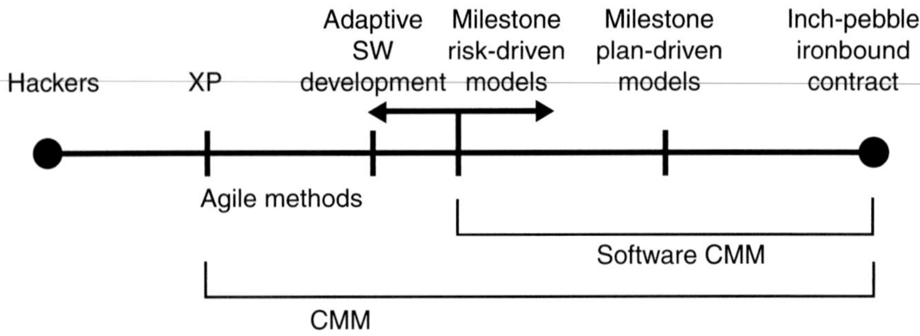

FIGURE 3.1 The planning spectrum. Unplanned and undisciplined hacking occupies the extreme left, while micromanaged milestone planning, also known as inch-pebble planning, occupies the extreme right. SOURCE: B. Boehm, "Get Ready for Agile Methods, with Care," *Computer* 35(1):64-69, January 2002.

release quality, which is the most important quality metric, along with general trends in the improvement of programmer productivity, customer satisfaction, and team morale.

Agile processes are fundamentally different from the practices adopted during traditional development, and represent much more than a mere time-compression of the waterfall development process. Several practices that are part of ASD are significantly different from the traditional waterfall software development process. For example, some of the more common ASD practices from XP and FDD are test-driven development, continuous integration, collective code ownership, and small releases.

As an example, consider one of the more popular ASD processes—Scrum. Work is structured in cycles of work called sprints, iterations of work that are typically 2 to 4 weeks in duration. During each sprint, teams pull from a prioritized list of customer requirements, called user stories, so that the features that are developed first are of the highest value to the customer. User stories are a good way for users (and other stakeholders) to express desired functionality. At the end of each sprint, a potentially shippable product is delivered.[17]

Scrum essentially entails building a system in small, shippable-product increments. The shorter release cycles allow systems to obtain early feedback on various aspects, such as usage profile, quality, dependencies, integration, and stress loads, from relevant stakeholders. This information can then be used to guide corrective action affordably during development. The development team decides the deliverable for each sprint with input from the end user. This allows the end user to understand the qual-

[17] See the Scrum Alliance homepage at http://www.scrumalliance.org.

TABLE 3.1 Results of Using Extreme Programming (XP) in Industrial Software Systems: Summary of Four Case Studies

Hypothesis: The use of a subset of XP practices leads to an improvement in:	IBM Case Study: Small, Co-located Systems	Sabre Airline Solutions Case Study: Small, Co-located Systems	Sabre Airline Solutions Case Study: Medium, Co-located Systems	Tekelec Case Study: Small, Distributed Systems
Pre-release quality	Yes	Yes	Similar-yes	N/A
Post-release quality	Yes	Yes	Yes	Yes
Programmer productivity	Yes	Yes	Similar-higher	Yes
Customer satisfaction	Yes	N/A	N/A	Neutral-satisfied
Team morale	Yes	N/A	N/A	N/A

NOTE: N/A, not available.

SOURCE: Created with data from Laurie Williams, William Krebs, Lucas Layman, and Annie I. Antón, "Toward a Framework for Evaluating Extreme Programming," *Proceedings of the 8th International Conference on Empirical Assessment in Software Engineering,* Edinburgh, Scotland, May 24-25, 2004, pp. 11-20; Lucas Layman, Laurie Williams, and Lynn Cunningham, "Motivations and Measurements in an Agile Case Study," *Journal of System Architecture* 52(11):654-667, 2006; Lucas Layman, Laurie Williams, and Lynn Cunningham, "Exploring Extreme Programming in Context: An Industrial Case Study," *Proceedings of the 2nd Agile Development Conference,* Salt Lake City, Utah, pp. 32-41, June 22-26, 2004; Lucas Layman, Laurie Williams, Daniela Damian, and Hynek Bures, "Essential Communication Practices for Extreme Programming in a Global Software Development Team," *Information and Software Technology* 48(9):781-794, September 2006.

ity of the system being delivered and to make changes to the requirements as the project proceeds. Within each sprint, engineers continue to use tested and proven development practices and development tools.

Other ASD processes such as XP prescribe in detail particular practices to be used and afford little flexibility, which might present a challenge to application in the DOD context. However, care should be taken when adopting some of the agile development processes. Some of them advocate, for example, not documenting code (more popularly known as self-documenting code) or not requiring any initial architecture work (such as XP). These practices would likely be catastrophic for large systems that are expected to endure. Nonetheless, even if specific prescriptions are adapted to the particular context, adopting IID based on ASD methodologies as the default DOD practice would require radical changes to current oversight and management processes.[18]

As an example of the use of ASD for defense systems, recent work by Crowe and Cloutier[19] presents the results of a DOD case study in which a phased approach was used to deploy high-priority capabilities using an incremental agile process. This approach was adopted for the Defense Readiness Reporting System-Army (DRRS-A). By 2006, the U.S. Army Readiness Reporting System was no longer meeting the needs of the commanders. In 9 months, a new system was built to meet these needs without the loss of existing capabilities by integrating all aspects of the software life cycle using an agile approach. The team involved about 60 people from government and several contracting firms. A development sprint length of 30 days was used. In addition to recognition of the importance of communication, coordination, and risk management, key lessons included the need for tight collaboration among the contracting teams, stakeholders, and program offices and ensuring by means of predefined checkpoints that the development met the customers' needs. All of these were viewed by the participants as significantly enabled by the agile development processes. Table 3.2 lists emergent characteristics of the agile process.

Large software development organizations make use of IID processes in a variety of ways, including for individual logical components of a more complex product set (Box 3.3). There can and will be multiple coding milestones and beta versions in a product release, with features constantly released to end users for feedback. This feedback is then leveraged to improve the quality of the end product iteratively. Within each beta or candidate release, development teams use the normal agile development practices (e.g., sprints) and focus on planning, work allocation, and quality practices such as unit testing and code inspections; each release is equivalent to an individual system release. This approach provides complete transparency to the development process and enables project managers to get an early indication of problems in development, slippage in meeting milestones, and quality issues.

The above descriptions are meant to give a flavor of what IID looks like in the context of various agile software methodologies. Although pure

---

[18] Don Johnson, Office of the Deputy Assistant Secretary of Defense for Networks and Information Integration, "Challenges in Acquisition of Information Technology," presentation to the committee, December 8, 2008.

[19] Portia Crowe and Robert Cloutier, "Evolutionary Capabilities Developed and Fielded in Nine Months," *Crosstalk* 22(4):15-17, May/June 2009.

TABLE 3.2 Emergent Characteristics of the Agile Process

Characteristic	Comments
Liberty to be dynamic	Agility needs dynamic processes while adhering to acquisition milestones.
Nonlinear, cyclical, and nonsequential	The life-cycle behavior was not like traditional waterfall models or linear frameworks; decreasing cycle times.
Adaptive	Conform to changes, such as capability and environment.
Simultaneous development of phase components	Rapid fielding time may not lend itself to traditional phase containment (i.e., training and software development together).
Ease of change	Culture shift to support change neutrality; ease of modification built in to architecture and design.
Short iterations	Prototyping, demonstrating, and testing can be done in short iterative cycles with a tight user-feedback loop.
Lightweight phase attributes	Heavy process reduction, such as milestone reviews, demonstrations, and risk management.

SOURCE: Portia Crowe and Robert Cloutier, "Evolutionary Capabilities Developed and Fielded in Nine Months," *CrossTalk* 22(4):15-17, May/June 2009.

agile methods such as Scrum or XP may not be appropriate for DOD software efforts (for example, the short time lines such as the 30-day sprints advocated in Scrum), the core concepts of the methods are nonetheless applicable to defense IT acquisition. Of particular importance is the idea that the IID cycle must constantly obtain and reflect end-user feedback, especially for software development and commercial off-the-shelf software integration (SDCI) IT programs, so as to acquire systems aligned with end-user expectations and needs. Although the precise IID template for commercial off-the-shelf hardware, software, and services (CHSS) programs is somewhat different, the basic concepts are equally applicable to those programs.

BOX 3.3
Longer-Term Development Processes Can Also Employ Iterative, Incremental Development

For some very large systems, such as the Windows operating system family, the typical development life cycle is long—on the order of 3 years. Nonetheless, agile development processes are used within that span. There is a usable version of the system at any given instant, and new versions with features integrated are produced at regular intervals/milestones. The development process provides for continuous feedback from end users. Each phase has a well-defined release feature set. User feedback drives increment planning and feature increment, and assessment continues through multiple cycles, culminating in final release.

The process components and steps shown in Figure 3.3.1 are explained as follows:

- In the *start* phase, the already-existing system (if any) is defined as a baseline and the feature set is discussed in conjunction with the users.
- At the end of the *P + RM* (planning and requirements milestone phase), most of the decisions regarding the development of various features across multiple release milestones are made.
- In the *Coding Phase 1,* a significant amount of the feature-independent code (if any) is implemented, in addition to actual features.
- A significant proportion of the features are made available for end-user assessment through a *beta* release.
- Applied feedback, along with remaining features, is released in the *candidate release.*
- Based on field feedback and updates addressing any other end-user concerns, the main *release* takes place.

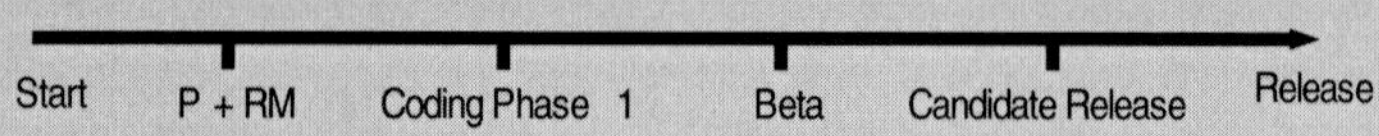

FIGURE 3.3.1 Development time line for some very large systems.

PLATFORMS AND VIRTUALIZATION: KEY UNDERPINNINGS FOR INFORMATION TECHNOLOGY SYSTEMS

The concepts of platforms and virtualization are key to modern software development and have important implications for DOD IT system acquisition.

A *platform*[20] is an evolving combination of hardware, system software, and applications software on top of which a wide group of individuals and organizations can innovate. Important examples today include Web 2.0 capabilities that enable interactive Web sites, the Windows family of operating systems, and the Intel x86 instruction set (implementations are also available from Advanced Micro Devices).[21]

Platforms can be defined at a low level in the technology stack (e.g., Ethernet for local area networking) or higher up. The higher up the stack one proceeds in defining a platform, the less commoditized and generic and the more application-, domain-, and environment-specific the platform becomes. Specifying a platform at higher layers of the stack for use across a portfolio of related IT programs has the advantage of establishing an architecture with a set of inherent characteristics that can then be taken advantage of by all of the capabilities built on top of the platform. The characteristics that can be established in such platform architecture include security and information assurance, operational availability, continuity of operations and disaster recovery, scalability, extensibility in provisioning, and extensibility in operations. These are all characteristics that are inherent attributes of an underlying architecture, and there is no reason that every IT program in a portfolio should be required to address them from the ground up. There is a trade-off, of course, between defining a platform with a rich set of services as described above and the inevitable filtering out of commercial off-the-shelf (COTS) (or other) offerings that can run on that platform. Such decisions would need to be made carefully.

Beyond the utility infrastructure layers, which are predominantly CHSS, there are several additional technology stack layers that one can consider incorporating into a platform, including middleware (plus integration code to make it suitable in a particular environment), common applications, and enterprise data repositories.

Middleware appropriate for the application domain plus any additional integration code necessary to make it function in the intended operational domain constitutes the next logical layer to consider in establishing a platform for use across a portfolio of SDCI IT programs. This, for example, could include COTS service-oriented architecture middleware plus the domain-specific integration code necessary to enable it to operate across bandwidth-limited, widely distributed environments such as those

---

[20] A. Gawer and M. Cussamano, *Platform Leadership*, Harvard Business School Press, Boston, Mass., 2002.

[21] See National Research Council, *Assessing the Impacts of Changes in the Information Technology R&D Ecosystem*, The National Academies Press, Washington, D.C., 2009, for a discussion of major platforms, their relationship to innovation, and their evolution over time.

on deployed units—for example, to provide reliable messaging or distributed security services in a deployed distributed environment. The COTS middleware products typically will not directly accommodate such very demanding environments without additional integration effort. Other software layers that can be considered for standardization in a platform include common applications (e.g., database and application servers) and key enterprise information repositories.

At the same time, however, adhering to the standards and design rules of the platform architecture does impose constraints on those building and innovating on the platform. The more specific a standard or a design rule becomes as one progresses up the technology stack, the more difficult and time-consuming it is to gain broad adoption, the more difficult and time-consuming it is to evolve it to keep pace with relentlessly advancing technology, and the more suboptimal its application is to new domains. There are risks from monoculture and from being locked into a single supplier as well. These considerations thus argue for restricting the applicability of a platform to a collection of similar application domains such as a portfolio of programs. Determining the appropriate granularity at which to make use of a common platform is important and should take into account all of these factors. Done appropriately in the context of a portfolio of programs, a platform can enable significantly increased efficiency, agility, and speed to capability.

Virtualization refers to the abstraction of computer resources. For example, multiple virtual servers can run on a single physical server, and a virtual private network creates a secure network link on top of an underlying physical network. Virtualization offers a number of benefits. It enables a portfolio of programs to deliver capability independent of hardware deployment and therefore increases efficiency, agility, and speed to capability. These increases provide ample reason for separating the COTS hardware components of an IT program from the software development and COTS software integration elements of an IT program. This is true both for infrastructure-based capabilities that have ample bandwidth to permit the use of centralized computing centers, and for deployed capabilities where bandwidth limitations often dictate a deployed version of a virtualized computing, storage, and network utility infrastructure.

Virtual platforms are of particular value in a deployed environment, where the resulting efficiencies can lower the lift capacity required to deploy a combat unit, as well as lower the space, weight, power, and cooling required in a ship or an aircraft installation. Virtual platforms also have the potential to lower sustainment costs otherwise inherent in program-specific approaches to logistics support. Finally, virtual platforms enable SDCI IT programs to focus exclusively on deployment decisions involving only their own specific systems as machine images, and at the same time to avoid addressing production decisions involving environmentally qualified COTS-based hardware installation.

A RECOMMENDED ACQUISITION MANAGEMENT APPROACH FOR INFORMATION TECHNOLOGY PROGRAMS

Chapters 1 and 2 of this report highlighted the difficulty of applying a governance and oversight regimen based on hardware and weapons system development to IT programs. Advocates of the current governance and oversight structure sometimes assert that this structure permits tailoring and provides all the flexibility needed for an MDA or a PM to adjust the way that the process is applied to specific programs. Even if these advocates are right, given the central importance of the rapid and effective acquisition of IT systems, there should be no reason that each MDA and each PM be required to define de novo a development and oversight process attuned to an IT program. Rather, as already argued in Chapter 2, a more effective approach is to establish a separate and distinct program governance and oversight regimen for IT programs that leverages the significant body of research available and the more than 20 years' worth of past recommendations and conforms to the widely adopted commercial best practices for development.

IT programs can be defined to focus primarily on the acquisition of developmental software, COTS software integration, COTS hardware, COTS software, commercially available services, or combinations of these. There are fundamentally different classes of issues involved in each of these cases. This examination focuses on the two categories of IT programs identified in the introductory chapter of this report:

- *SDCI programs*—those focused on the development of new software to provide new functionality or focused on the development of software to integrate COTS components, and
- *CHSS programs*—those focused exclusively on COTS hardware, software, or services without modification for DOD purposes (i.e., the capabilities being purchased are determined solely by the marketplace and not by the DOD).

The hardware components of information technology programs are most heavily influenced by Moore's law, which predicts the doubling of capacity per unit expenditure every 18 months. The resulting advance of networking, computing, and storage capacity in COTS hardware led the DOD to begin abandoning purpose-built military hardware and to embrace COTS hardware for IT programs beginning in the 1980s, even for tactical systems. For example, in the Navy, the Desktop Tactical Computer I (DTCI; a Hewlett-Packard 9020) was procured beginning in 1984 for various programs, including the Joint Operational and Tactical System I, the Integrated Carrier Antisubmarine Warfare Protection System, the Submarine Force Mission Planning Library, and numerous other small programs. The DTC I was followed in 1989 by the DTC II, based on the Sun Series 4/110 (later the Series 4/300), which was also ruggedized for shipboard use to be survivable under strenuous environmental conditions including shock and vibration. This change took place against the backdrop of commercial technology migrating from a data-center-based computing environment to a highly networked, client-server environment with substantial computational power available at every desk.

The trend toward COTS was further reinforced in 1994 when Secretary of Defense William Perry issued a memorandum prohibiting the use of most military standards without waivers and encouraging in their place the use of performance specifications and industry standards.[22] In this commercial market environment dominated by Moore's law and a high rate of technology change, hardware obsolescence and supportability have become issues that IT programs have to deal with.

In contrast to the hardware components, the software components of IT programs are most heavily influenced by the fast pace of technology change in the Internet environment and by the fundamental difficulty in defining requirements[23] for many classes of software systems, especially (but not limited to) human-interactive software systems. In the commercial Internet environment that defines the experience base of most young men and women entering the military today, new software capabilities are introduced on a regular and routine basis, often with significant end-user involvement and feedback through early alpha and beta release programs.

Fundamentally different classes of issues are involved when dealing with hardware versus software components of IT programs. Therefore, different strategies are appropriate when addressing the different components. In both cases, rapid change is a fundamental factor that must be addressed, and IID acquisition strategies are indeed appropriate. However, the nature of the capability increments should differ for hardware and software components owing to the different issues driving them.

Although some DOD IT programs are defined as exclusively COTS hardware acquisition programs (e.g., the acquisition of networking, com-

[22] William Perry, Memorandum from the Secretary of Defense to the Secretaries of the Military Departments, "*Specifications and Standards—A New Way of Doing Business*," June 29, 1994, Office of the Secretary of Defense.

[23] F.P. Brooks, Jr., "No Silver Bullet—Essence and Accidents of Software Engineering," *Information Processing* 20(4):10-19, April 1987.

puting, or storage infrastructure) or exclusively software development acquisition programs, others are defined to provide both software-based capabilities and the physical hardware plant on which the software will run. This has been particularly true for IT programs defined to provide deployable capability such as command, control, communications, computers, intelligence, surveillance, and reconnaissance (C4ISR) or combat support programs for deployed land, air, and maritime force elements. During the early days of COTS hardware adoption by the DOD, the capacity of available hardware and the nature of the software environment dictated that many systems be single-purpose.

Several key factors have changed significantly since those early days, however. The capacity available in commodity COTS hardware has grown exponentially, following Moore's law. Software is becoming increasingly independent of hardware in many DOD IT programs. At the same time, operations and sustainment costs have grown along with the complexity of IT systems. In addition, virtualization technology has matured. This has created a compelling business case for virtualized computing, storage, and network infrastructure utility models in many enterprise environments. The existence of such infrastructure can decouple the time cycles of SDCI capability increments from the need to deploy or refresh hardware, allowing SDCI IT programs or portfolios of programs to deploy capability to end users with significantly increased speed and agility. This is particularly true for server applications. With regard to client-side components, the use of a standardized technology platform on top of a virtualized computing, storage, and network infrastructure can further increase speed and agility.

For all of these reasons, IT programs, even those intended to provide deployed or deployable capability, should be defined as exclusively SDCI IT programs or as exclusively CHSS IT systems programs. Absent a compelling case to the contrary, the CHSS components of IT systems and SDCI software elements of IT systems should be acquired through independent IID acquisition programs with the nature of the capability increments structured to address the different issues associated with hardware and software components. Moreover, platform-based virtualized computing, storage, and networking utility models should be favored wherever possible.

The following subsections offer a modified acquisition management approach recommended by the committee for SDCI IT programs and for CHSS IT programs. They also further define the use of a platform- and virtualization-based approach and show how this approach can be visualized and managed as a combination of these two categories of programs.

Proposed Acquisition Management for SDCI Programs

IT programs that are focusing on SDCI efforts are most heavily influenced by elevated end-user expectations that are based on the pace of technological change experienced by most users every day in the Internet environment, and by the fundamental difficulty in defining requirements for many classes of software systems, especially human-interactive software systems. SDCI IT programs must therefore focus not only on the functional and nonfunctional capabilities that they are chartered to provide, but also on employing IID practices for software development and program management practices that represent the best practices in software engineering.

As discussed earlier in this chapter, the roots of IID software development methods can be traced back many years. Nevertheless, the influence of the waterfall model persists to this day in the current DOD governance and oversight process. One possible reason for this persistence identified by Curtis and co-authors[24] is that major elements of the document-intensive waterfall method attempt to satisfy management's goals for accountability despite the fact that many of these elements fail to account for the successful execution of IT projects. This type of effort places increasing focus on the acquisition process at the expense of focus on the product. To break this hammerlock on software-intensive IT programs, several aspects of the program structure and the governance and oversight regimen would need to be changed through the adoption of the following core principles:

- Emphasis on shorter cycle times to deliver the best IT to the warfighter
  - —Time-boxed incremental deliveries of usable capabilities (also known as capability increments).
  - —Time-boxed iterations within each capability increment.
  - —Early focus on nonfunctional requirements and an architecture suited for the intended operating environment.
- Streamlined processes for requirements definition, budgeting, operational testing, and oversight
  - —Focus on "big-R" requirements[25] during early planning.

[24] W. Curtis, H. Krasner, V. Shen, and N. Iscoe, "On Building Software Process Models Under the Lamppost," pp. 96-103 in *Proceedings of the International Conference on Software Engineering*, IEEE Computer Society Press, Los Alamitos, Calif., 1987.

[25] "Big-R" requirements convey a widely recognized purpose, mission, and expected outcome (for example, a missile system would be assessed on the basis of its ability to hit a target at a given range under certain specified conditions). "Small-r" requirements involve a set of more detailed requirements associated with specific user interfaces and utilities

—Performance of integrated testing and evaluation commensurate with risk and benefit.

- The employment of IID methods for development, contracting, and testing
  —In particular, the voice of the user as a prominent factor throughout each iteration within each capability increment.
  —An acquisition governance process that empowers end users in the acquisition oversight decision processes.
- The decomposition of larger programs into smaller projects or increments that are delivered to the user in an evolutionary manner
  —Deployment decisions driven by risk and benefit.
  —Incremental build-out of the architecture in scope and scale sufficient to meet the needs of the functional requirements of each capability increment.
  —Long-term stable funding across multiple capability increments.

To fully internalize these core principles, SDCI programs should be structured as IID programs with time-boxed capability increments of not longer than 12 to 18 months to deliver meaningful capability to end users. The recommended acquisition management approach for SDCI IT systems programs is shown in Figure 3.2. Key to IID software methods is the acknowledgment of the difficulty of document-centric attempts to fully specify requirements in advance and the necessity of replacing such an approach with an iterative, incremental learning and communications process taking place between developers and end users. The capability increments are time-boxed and further broken down into a sequence of time-boxed iterations, each of which results in integrated and tested products with the voice of the end user further refining and reprioritizing the more detailed "small-r" requirements at the completion of each iteration. Each iteration will include analysis, design, development, integration, and testing to produce a progressively more defined and capable, fully integrated and tested product. This process is shown in Figure 3.3 as an adaptation of the traditional systems engineering "vee-diagram."

Each iteration would take on a subset of the overall problem of building the desired capability for the increment and would perform the full set of tasks, including requirements analysis and refinement, architecture formulation or refinement, design formulation or refinement, implementation, integration, and testing. It is natural that early increments would spend more time in the front-end processes of architecture and design, while later increments would spend more time in the back-end processes

---

that will evolve within the broader specified architecture as articulated in the initial big-R requirements.

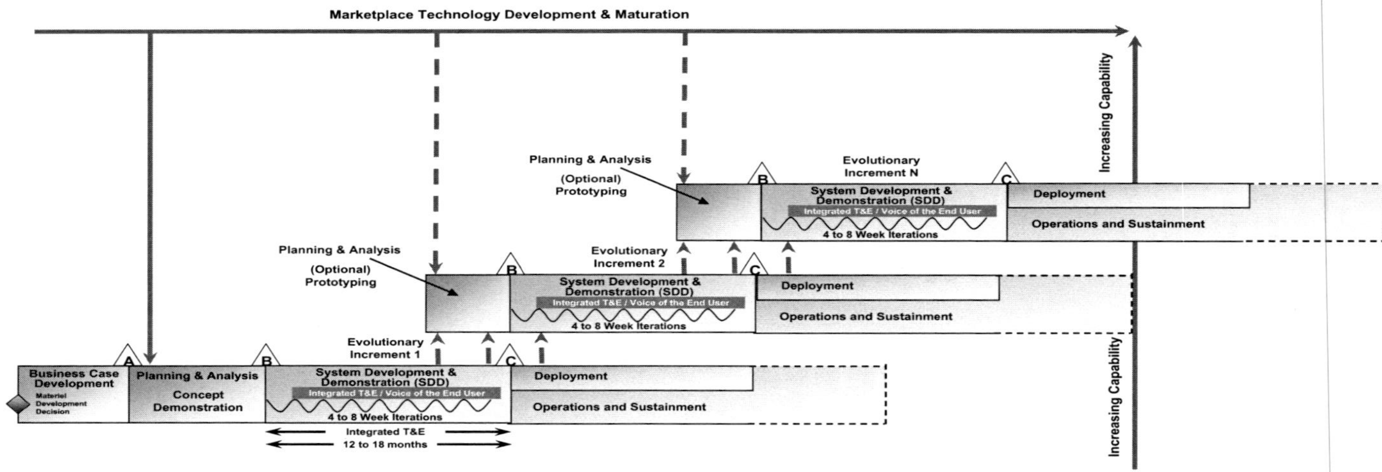

FIGURE 3.2 The acquisition management approach recommended by the committee for software development and commercial off-the-shelf integration (SDCI) information technology system programs. NOTE: T&E, test and evaluation.

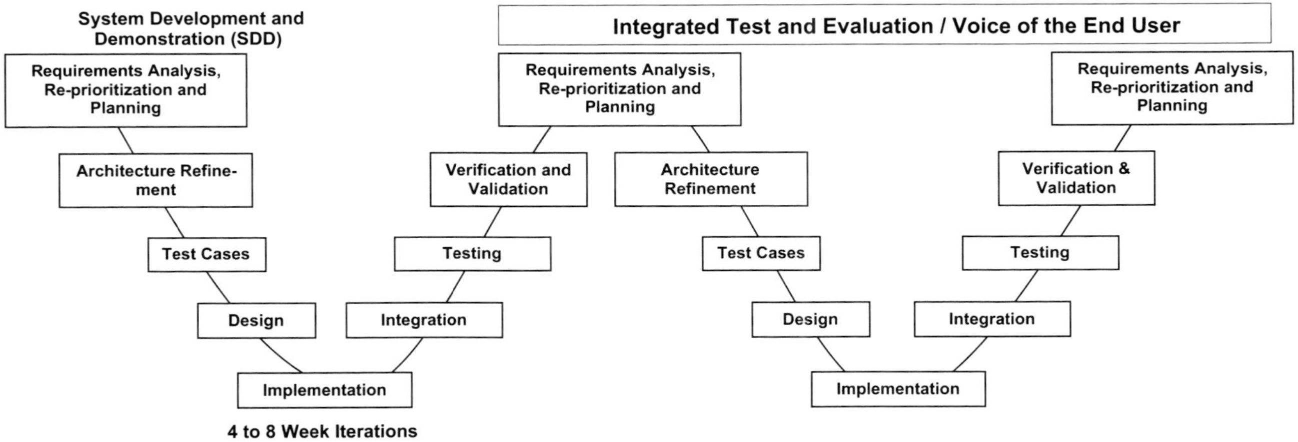

FIGURE 3.3 Time-boxed iterations within each capability increment.

of integration and testing. The two steps in the iteration of integration and testing are essential elements, because they serve to force direct understanding of the progress (or lack of progress) actually being made at frequent, well-defined checkpoints.

From one perspective this approach can be regarded as a traditional application of systems engineering processes, but there are several important differences. As just mentioned, testing plays a key role not only in terms of how often and when it is conducted, but also in how it is considered early in each iteration and the manner in which the "voice of the end user" is integrated into the process. Many regard some forms of IID as a test-driven process, since testing is addressed and tests are defined even before design and implementation. Further, when testing is actually conducted, the voice of the end user is integrated directly into the process, with direct effects on the further refinement of requirements and on the planning, and potentially the reprioritization, of subsequent iterations. This sequence of events reflects both a learning cycle and a communications cycle in each iteration.

With regard to integrated test and evaluation (T&E), operational end users are, of course, not the only important stakeholders. Other stakeholders should also be involved in the process, including developmental and operational test stakeholders, security certification and accreditation stakeholders, and interoperability stakeholders. They must all actively participate in testing and a nontraditional form of verification and validation (V&V)—not the traditional version of V&V against a fixed, predefined set of requirements, but rather the learning and communications process of IID that acknowledges the impossibility of complete up-front specification and the need to refine requirements as learning occurs. For such an approach to succeed and materially change the dynamics of the current time-consuming and unsatisfying process, all of the stakeholders that warrant a voice in the ultimate fielding decision should be integrated into this voice-of-the-end-user learning and communications cycle. A case can certainly be made that the operational end user is the most significant among the various entities that have an interest in the outcome and that the voice of end users should therefore be louder than that of the others. Chapter 4 presents more detail on testing and acceptance and on the formulation of an acceptance team that can serve as and also channel the voice of the end user.

The content of early iterations should be focused on a combination of the most technically challenging elements of the capability increment and on functional capabilities with the greatest business or warfighting value. The voice of the end user should provide feedback on each iteration for the refining and prioritizing of requirements in order to institutionalize the learning and communications process vital to IID. Since this voice of the end user is to play such a prominent role in refining and prioritizing requirements at each iteration within each capability increment, the requirements allocated to each increment should shift from the current focus on detailed functional requirements to a greater focus on objectives or big-R requirements. This implies a profound change to the Joint Capabilities Integration Development System process, for example, which is currently biased toward the document-centric, up-front detailed specification of functional requirements.

Nonfunctional requirements for software-intensive IT systems such as security and information assurance, operational availability, scalability, and performance are fundamental attributes of systems architecture, especially in distributed systems, and they need to be communicated clearly by the appropriate stakeholders. Similarly, the operational environment in which such a system must function can profoundly affect the system architecture. For SDCI programs of any significant scope and scale, system architecture is likely to represent one of the most significant risks, and it should be addressed early in the program, even during the concept development phase in advance of entering the development of capability increments. Here again, however, caution and pragmatism must be exercised to prevent susceptibility to the demand that all requirements must be fully documented up front.

Although these nonfunctional and environmental requirements will have a profound effect on the system architecture, building out the system architecture is itself a learning and communications process that is best accomplished in an IID fashion. Similarly, although one can perhaps understand the top-level nonfunctional requirements very well up front, their refinement into lower-level requirements that drive design and implementation is also best addressed in an IID fashion. At the same time, this should not be construed to suggest that big-R requirements will not change. Feedback on those is also important and should be accommodated.[26]

Any SDCI program must have a clear vision of an end-state target architecture appropriate to support both the full scope of the intended capability and the full scale of the intended deployment from the outset. It is at a minimum impractical and in most cases impossible to implement or even to fully prove the full scope and scale of the architecture either in the early pre-Milestone B concept development phase or in early capabil-

[26] IEEE, ANSI/IEEE Standard 1471-2000: Recommended Practice for Architectural Description of Software-Intensive Systems, October 2000; Philippe B. Kruchten. "The '4+1' View Model of Architecture," *IEEE Software* 28(11):42-50, November 1995; Mark W. Maier and Eberhard Rechtin, *The Art of Systems Architecting*, 2nd Edition, CRC Press, Boca Raton, Fla., 2000.

ity increments or iterations. The architecture, like the other aspects of the software in an IID approach, should be built up in scope to support the planned functionality of each capability increment, and should be built up in scale to support the planned deployment scope of each capability increment. Any attempt to force more up-front architecture proof or development will substantially delay a program's ability to deploy useful capability to end users.

Since the voice of the end user can adapt priorities across iterations within each capability increment, the learning that takes place through this process can affect the ability to achieve all of the objectives originally targeted for the increment. Further, the continuous nature of T&E inherent in IID and the learning and communications process integral to IID will develop substantial T&E results as the iterations progress within a capability increment. Therefore, an integrated approach to T&E to include the voice of the end user; traditional development, testing, and evaluation; operation testing and evaluation; interoperability certification; and information assurance certification and accreditation equities is a fundamental element of this modified acquisition management approach for IT programs. As was the case with the requirements process, this implies a profound change in the T&E process used for such programs. At the conclusion of each time-boxed capability increment, the focus of the deployment decision (Milestone C) shifts from the evaluation of successful completion of the predefined scope allocated to the capability increment to an evaluation of the risks and benefits of deployment combined with a reevaluation of the priorities for the objectives of the subsequent capability increment (Milestone B). This integrated approach to test and evaluation is discussed in depth in Chapter 4.

The governance and oversight model for such an IID program must change substantially from the current structure to accommodate this approach. The vast majority of SDCI IT system programs use technology developed and matured in the commercial marketplace. While these programs do develop software, they do not develop fundamental technology. Rather than focusing on technology development and technology readiness levels (TRLs) and technology readiness assessments (TRAs), the early phase of the program structure should use prototyping to demonstrate key concepts that the system is intended to address, key nonfunctional requirements, high-business-or-warfighting-value functional requirements, and an ability to function properly in the intended operational environment for the system, including constrained communications and networking environments. Once these issues are resolved, there is no need to repeat this redefined concept development phase for each capability increment. All that is required for successive capability increments after the first is a much-abbreviated planning and analysis phase to reevaluate the plan for the next increment, given the learning that took place and the results actually achieved in the previous increment.

From a program oversight and governance perspective in the current DODI 5000 approach, there can be multiple oversight bodies, and there is a large number of participants in the program oversight and review process. Each of the oversight groups can require program changes, and, often, each individual representative of a Service or other group has the ability to force changes to a project or require special accommodations or requirements at a very detailed level, often without any justification of the impacts on cost or schedule caused by the changes. This has negative effects: (1) there are too many requirements; (2) the program is not able to effectively prioritize requirements; and (3) the requirements can in fact be contradictory or extremely difficult to implement.

The responsibility, authority, and accountability for program execution all need to be clarified and strengthened. Program authority and accountability are diluted by the scope and complexity of the programs and the resulting program structures. Further, the January 2006 report of the Defense Acquisition Performance Assessment (DAPA) project concluded that "the budget, acquisition and requirements processes [of the Department of Defense] are not connected organizationally at any level below the Deputy Secretary of Defense."[27]

To address these issues, the program manager and portfolio management team (PMT), as defined in Chapter 2, should have decision authority, derived explicitly from higher authority, to determine trade-offs among schedule, cost, and functionality. One of the most important tasks of the PM and PMT is to determine priorities in the IID development: what is in the first capability increment, what is in the second, and so on. Further, the PM and PMT decide which requirements are essential and which would be "nice to have." The PMT's role is vital to ensuring that appropriate interoperability is maintained across the team's functional area.

This process is not workable if there is program oversight or accountability to one or more committees, with each member being able to force changes. In terms of functionality, such committees have a constructive role in offering their views, but the PM and PMT should derive authority from a higher source: the milestone decision authority. (In some cases it may be appropriate for the PMT to be delegated the role of MDA.) Having governance committees be strictly advisory, with decision authority clearly given exclusively to the PM, PMT, and MDA as discussed below for each program phase and decision milestone, is key to the success of

[27] Assessment Panel of the Defense Acquisition Performance Assessment Project. *Defense Acquisition Performance Assessment Report*, Department of Defense, Washington, D.C., January 2006.

the committee's recommended approach. Of course, with greater authority must come greater accountability. The PM and PMT are accountable to a senior DOD official in the chain of command—namely, the MDA, who in turn is accountable for the success of the program.

Another important element of incremental program management is that the PM and PMT have an effective body to help them make the right decisions on priorities and specific requirements for each stage of the project. As indicated above, this would be an advisory body, and the decision responsibility would reside with the PM and PMT. The user has to have a significant voice here.

Finally, since multiple time-boxed capability increments will fit within each budget cycle of the planning, programming, budgeting, and execution process, and to give end users confidence that their requirements will be addressed (thereby avoiding the unintended but real consequence of users trying to overload their requirements into the first capability increment), IID programs should be provided with a stable budget profile across multiple capability increments.

Appendix B provides a more detailed discussion of each of the program phases and decision milestones presented in Figure 3.2. For each decision milestone, the key objectives of the milestone and the responsibilities of the PM, PMT, and MDA are addressed. (Once again, note that in some cases it may be appropriate for the PMT to be delegated the role of MDA.)

Proposed Acquisition Management for CHSS Programs

The goal of CHSS programs is to exploit commercially available products and services without modification to meet DOD needs, although ruggedization to meet environmental requirements for deployed or deployable systems can be addressed in this category of programs. Key requirements that should be addressed in such programs include capability, capacity, scalability, operational availability, information assurance, and, in the case of deployed or deployable programs acquiring hardware, the environmental qualification of the hardware components and any associated ruggedization. Products in this category are often relatively interchangeable components, with performance characteristics that are completely understandable through specifications or product data sheets and through inspection or experimentation to validate vendor claims. (Therefore, there is no real reason for confusion about the basic capability or service procured, and the areas of differentiation between competing alternatives are entirely knowable and verifiable.)

A key characteristic for commercial components and services is the rapid pace of change driven by the marketplace. For more than three decades, IT hardware in the commercial marketplace has been driven by Moore's law, which describes the approximate doubling of capacity per unit of expenditure every 18 months. Although that growth in performance cannot continue indefinitely, IT capacity will continue to improve, and such improvements need to be factored in to the acquisition approach of IT systems programs providing COTS hardware. COTS software is most heavily influenced by the fast pace of technology change in the Internet environment, often driving version upgrades every 12 months or less. Even the most complex COTS software products on the market, which may have 2-year cycles between major version updates, typically have several minor version updates between the major updates, with the minor updates on a cycle of 12 months or less. Finally, one must take into account the case of IT services—capabilities offered over the network on an ongoing basis, as opposed to localized applications. Such services are typically offered based on COTS hardware plus COTS software plus an IT service delivery and management process. To remain competitive in the marketplace, service providers too must regularly refresh the COTS hardware and software providing the basis for their service, as well as innovate in the COTS service delivery and management dimension.

Since the rapid pace of change driven by the commercial marketplace is such a driving factor for this category of IT programs, the structure of the programs should explicitly take this into account through an IID-based acquisition approach with iteration cycles of relatively short duration (18 to 24 month), just as was the case with SDCI programs.

For CHSS programs, this is as much a business strategy issue as it is a technical strategy issue, and both strategies should be addressed and vetted early in the program and revalidated as capability increments proceed. If appropriate virtualization and storage strategies are adopted to enable easy extension of software capability, then this utility also has to provide functional support in at least two other areas: provisioning and operations. The virtualized computing and storage utility model should provide not only the necessary capacity to support new and emergent applications, but also a means for operations staff to provision new virtual environments for those applications, to monitor the status of applications in operations, and to initiate corrective action in the event of abnormal behavior.

The capability increments for such a CHSS program should be driven by a combination of affordable investment profiles, technology refresh objectives, avoidance of technological obsolescence, and the time that it takes for installation across the production inventory objective for the program. Although Moore's law has historically operated on an 18-month time cycle, the useful life of networking, computing, and storage hardware is at least two to three times that duration. Additionally, if the IT program has a significant production inventory objective, it will take time to complete installation on all target units in the inventory objective, especially if shipyard or aviation depot facilities are required to accomplish the installations. At the same time, however, hardware will become increasingly difficult to maintain as it ages, as vendor support diminishes, and as it becomes harder to buy spares. If a virtualized computing and storage utility model is being employed as described earlier, avoiding technological obsolescence and providing the growth capacity to support a more rapidly changing collection of software applications produced by other IT programs are also counteracting forces. This leads to an IID-based acquisition model with increments driven by the roughly 18-month time cycle of Moore's law, a production rollout schedule driven by affordable investment profiles and the availability of target units in the inventory objective to accomplish installations, and a technology refresh across the inventory objective also driven by affordable investment profiles. In all cases the CHSS acquisition model should adhere to the philosophy of deploying "then year" technology in all new or upgraded production installations—that is, the model should use the production baseline established in the most recent IID capability increment.

For example, the combination of the sustainable investment profile and the availability of deploying units may dictate a deployment schedule of 4 years or longer for equipping the full inventory objective. At the same time, it would not be prudent to install technology that is more than 4 years old on the last group of units to be equipped. For hardware, it would make more sense to update the technology incrementally and to requalify the updated equipment every 18 months. This could lead to a strategy of deploying increment 1 of a capability to a third of the inventory objective in the first 18 months of the program, increment 2 to the second third of the inventory objective in the second 18 months of the program, and increment 3 to the final third of the inventory objective in the third 18 months of the program, and then initiating technology refresh on the first third of the inventory objective with an increment 4 in the fourth 18 months of the program, and so on.[28]

The structure of a CHSS IT program is as much a question of investment and business strategy as it is a question of technical strategy; all of these topics should be addressed early in the program and revalidated as capability increments proceed. The governance and oversight model for such an IID-based IT program can be substantially simpler than the current one described in Chapter 1. The technology development phase

[28] Admittedly, this technology refresh strategy raises challenges for those personnel responsible for operating and maintaining different generations of hardware. Such issues need to be taken into account when considering life-cycle costs during the development of the business case.

of the current regimen, with its focus on TRAs and TRLs and technology risk, is not applicable to COTS hardware-focused IT programs. Such an IT program would not be attempting to push the state of the art in information technology and would be using mainstream, or even commodity, COTS hardware and software. The Milestone A and B decisions that are at either end of the technology development phase can readily be combined and accomplished together. The engineering and manufacturing development phase can likewise be simplified. For COTS software programs, it becomes largely a COTS configuration effort. For COTS hardware programs, it becomes largely a COTS hardware integration effort or, at most, a ruggedization effort. This effort would be aimed at the environmental qualification of the COTS hardware baseline for survivability in the target production environment, followed by environmental qualification testing to arrive at a qualified hardware component. Such efforts may address factors such as shock, vibration, high or low temperature extremes, blowing sand, high humidity, and other environmental factors significant in the target operating environment of the IT system, especially for systems destined for deployed or deployable units.

Developmental and operational testing and evaluation for such an IT program can similarly be simplified to focus on validating the facts that capacity objectives have been met, that environmental qualifications have been achieved, that provisioning and operations support functions are effective, and that operational availability and other integrated logistics support objectives have been achieved. The vast majority of the operational testing and evaluation can be accomplished outside an operational environment, with only a final verification test taking place on an operational unit in a real-world operational environment. The resulting overall governance and oversight structure for such a COTS hardware-based IT program is shown in Figure 3.4.

Appendix C provides a more detailed discussion of each of the program phases and decision milestones presented in Figure 3.4. For each decision milestone, the key objectives of the milestone and the responsibilities of the PM, PMT, and MDA are addressed. The discussion in Appendix C focuses on the differences between this category of programs and the SDCI category discussed previously.

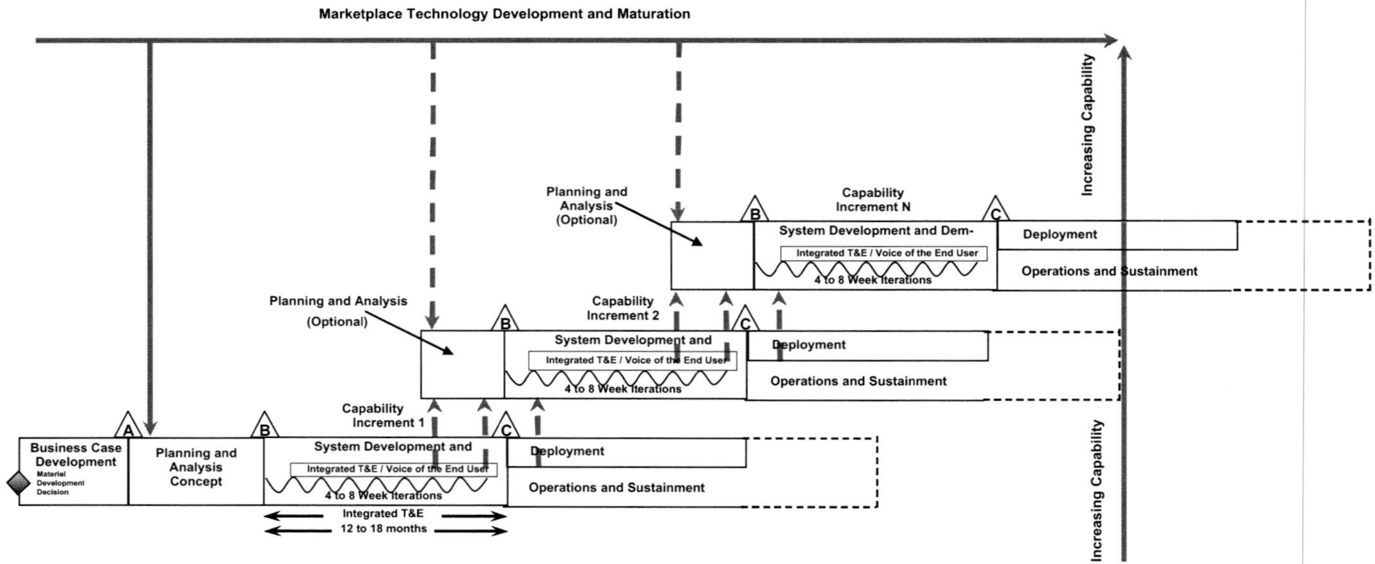

FIGURE 3.4 The acquisition management approach recommended by the committee for commercial off-the-shelf hardware, software, and services (CHSS) information technology programs. NOTE: T&E, test and evaluation.

4

Acceptance and Testing

INTRODUCTION

The test and evaluation (T&E) methodologies associated with weapons systems are mature and largely stable. In contrast, T&E methodologies for information technology (IT) systems are still maturing. Areas where challenges exist for IT systems include the assessment of Department of Defense (DOD) enterprise-level scalability, the proper role of modeling, cross-system integration validation, and interoperability validation. These and other areas lack widely agreed on test methods and standards, and they lack operationally realistic T&E methodologies.[1] Not surprisingly, commercial developers of large-scale applications experience similar challenges.

The tenets of DOD Instruction (DODI) 5000 have evolved over many decades and have served as the basis for well-established criteria and processes for decisions on weapons systems programs—for example, decisions on whether to enter into low-rate initial production or full-rate production—as well as providing commonly understood methods that comply with requisite policy and statutory guidelines. The equivalent

---

[1] See David Castellano, "Sharing Lessons Learned Based on Systemic Program Findings," presented at 2007 ITEA Annual International Symposium, November 12-15, 2007; National Research Council, *Testing of Defense Systems in an Evolutionary Acquisition Environment*, The National Academies Press, Washington, D.C., 2006; and Software Program Managers Network (SPMN), "Lessons Learned—Current Problems," *SPMN Software Development Bulletin #3*, December 31, 1998.

decision points in DOD IT systems are quite different in the iterative, incremental development (IID) processes discussed in Chapter 3. As a result, an equivalent understanding of what is required and when it is required has not been reached for IT systems acquisition. The results are frustration for developers and other participants in the acquisition process and uncertainty and delay in the process itself. Much can be gleaned from the experience of commercial IT systems developers and suppliers—such insights are just beginning to be incorporated into DOD practice. This chapter briefly reviews key elements and shortcomings of current practice and outlines opportunities for improvement, with a focus on making the perspective of the end user more salient.

SHORTCOMINGS OF PRESENT DEFENSE TEST AND EVALUATION

The current DOD process for the acquisition of IT systems has its roots in the procurement of hardware-oriented systems that will be manufactured in quantity. As such, the DOD's typical practice is to determine whether or not a design is adequate for its purpose before committing to advancing the production decision. For programs that are dominated by manufacturing cost, this approach reduces the possibility that a costly reworking of a system might become necessary should a defect be identified only after fielding of units in the operational environment has begun.

DODI 5000 directs programs to conduct testing and evaluation against a predetermined set of goals or requirements. As shown in Table 4.1, the acquisition process, including the T&E process, is governed by a large set of rules, test agents, and conditions, each trying to satisfy a different customer. Traditional test and acceptance encompass three basic phases: developmental test and evaluation (DT&E; see Box 4.1), the obtaining of the necessary certification and accreditation (C&A), and operational test and evaluation (OT&E; see Box 4.2).

In essence, the current approach encourages delayed testing for the assessment of the acceptability of an IT system and of whether it satisfies user expectations. A final operational test is convened in order to validate the suitability and effectiveness of the system envisioned as the ultimate deliverable, according to a specified and approved requirements document. This approach would work if the stakeholders (program manager [PM], user, and tester) were to share a common understanding of the system's requirements. However, because of the length of time that it takes an IT system to reach a mature and stable test, what the user originally sought is often not what is currently needed. Thus, unless a responsive process had been put in place to update the requirement and associated

TABLE 4.1 Test and Evaluation Activity Matrix

Activity	Test Agent	Conditions	Customer	Reference
Developmental Test and Evaluation (DT&E)	Program management office (PMO) and/or contractor and/or government developmental test organization	Determined by PMO; generally benign, laboratory; developer personnel	PMO	Title 10, DOD Instruction (DODI) 5000 series
Operational Test and Evaluation (OT&E)	Independent operational test agent	Operationally realistic, typical users	PMO, end user, and Milestone Decision Authority	Title 10, DOD Instruction (DODI) 5000 series[a]
Joint Interoperability Test Certification	Joint Interoperability Test Center	Applicable capability environments	Command, Control and Communications Systems Directorate (J6); the Joint Staff	DOD Directive 4630.5 DODI 4630.08 Chairman of the Joint Chiefs of Staff Instruction 6212.01D
Defense Information Assurance Certification and Accreditation Process (DIACAP)	Operational test agent, Defense Information Systems Agency (DISA), DISA Field Security Operations Division, National Security Agency	Operational, laboratory	Designated accrediting authority	DODI 8510.01[b]

[a] Also the Director of Operational Test & Evaluation (DOT&E) policy on testing software-intensive systems.
[b] Note also the DOT&E policy on information assurance testing during OT&E.

BOX 4.1
Developmental Testing

The process for testing U.S. military systems, including information technology systems, begins at the component level and progresses to the subsystem and system levels. Initial testing is done with components and subsystems in the laboratory, after which the testing graduates to larger subsystems and full systems.

Early developmental testing, which is conducted by the developer, is designed to create knowledge about the proposed system and about the best solutions to technical difficulties that must be confronted. Later, the Department of Defense (DOD) may participate in developmental testing that is designed to challenge the system under a wider variety of test conditions or variables. Next, more operationally realistic testing is conducted and overseen by the DOD.

BOX 4.2
Operational Assessments

Typically, operational testing, which incorporates user feedback, is more operationally realistic and stressing than is earlier developmental testing. Also, the Department of Defense may conduct operational assessments designed to measure the readiness of the system to proceed to operational testing. Overall, program success is ensured by ironing out performance issues in early operational assessments or limited user tests first.

These limited user tests are still developmental in nature, but they are designed to evaluate the military effectiveness and suitability of systems in a somewhat more operationally representative setting. For example, while still in development, information technology (IT) systems might be tested in configurations that provide interfaces with other related or ancillary IT systems on which development has already been completed. In addition, IT systems might be tested in environments that simulate the expected battlefield environments, which might involve shock, vibration, rough handling, rain or maritime conditions, and/or temperature extremes.

As with other testing, operational assessments are designed to provide data on the performance of a system relative to earlier prototypes, some of which might already have been deployed. These assessments are also designed to compare the rate of failures, system crashes, or alarms with those of earlier prototypes.

test plan, the tester could well be compelled to perform T&E against a requirement acknowledged by the time of the test to be obsolete but that once was needed, fully vetted, and approved.

Also, the DOD has not adopted a norm of requiring continuous user involvement in developmental testing. Obtaining the perspective of typical users early on and sustaining that user input throughout a development project are essential if the DOD is to exploit iterative, incremental development methods fully.

This situation is exacerbated, in part, because the DOD acquisition community has not been properly and consistently trained to understand IT systems and the development and integration processes of associated emerging systems. Mismatches between the approaches of developmental systems and the expectations of test systems will inevitably lead to failed or unhelpful operational testing. This mismatching manifests itself in at least two ways: in terms of technological expectations and of user expectations. If developers and testers fail to agree on which systems capabilities are to be supplied by the program versus those that are to be provided by existing systems, both developers and testers collectively fail because of mismatched test scope. If users fail to be engaged early in development or if testing fails to acknowledge the continuous revision of user-interface requirements, the development and testing processes collectively fail because of mismatched user engagements and expectations.

A lack of user engagement limits understanding by developers and testers as to what is required: for developers this means what capabilities to build, and for testers it means what capabilities to evaluate. In either case, neither party has an adequate opportunity to implement any corrective action under the current acquisition process.

Operational testers, independent evaluators, and the line organizations that represent end users are the key participants in operational tests. These tests include operationally realistic conditions that are designed to permit the collection of measurable data for evaluating whether a system is operationally suitable and operationally effective as measured against key performance parameters (KPPs). KPPs are descriptions of the missions that the system must be able to perform. Historically, meeting (or not meeting) the KPPs has been a "go/no-go" criterion for system fielding. By definition, a KPP is a performance parameter that, if not met, is considered by the DOD to be disqualifying. There have been many cases over the past decade or so in which the Service test community has documented "capabilities and limitations" in operational tests of systems developed in response to urgent operational needs statements from combatant commanders. This type of case provides more flexibility to decision makers, allowing them to decide whether a system is "good enough" to meet immediate wartime needs.

In most cases the units participating in the tests necessarily are surrogates for the full range of actual end users who will ultimately receive the fielded system. Their role is to be the representatives of the users and to bring to bear the users' perspective from an operational standpoint. Naturally, the independent operational test agency and independent evaluators will assess the ability of the system to meet the KPPs in the requirements document along with the strengths and weaknesses of the system. The operational test community will also assess whether the KPPs may or may not be the best contemporary measure of system acceptability at the time that the test is completed. Typically, operational testers, evaluators, and participating units have recent operational experience in the same or associated mission areas for supporting their assessments. Developers need similar operational insights, as the systems being developed must perform under realistic battlefield conditions and not just in the laboratory. The bottom line is that there is no substitute for a user perspective throughout the acquisition of IT systems.

Cost and schedule, rather than performance, frequently become the main drivers of a program, and developmental testing is too often given short shrift, or the amount of time allowed for operational testing is reduced. In general, program offices tend to sacrifice rigor in favor of a more condensed, "success"-oriented testing approach. As a result, user issues that should have been discovered and addressed during DT&E may escape notice until OT&E. Not only are problems more difficult and more expensive to fix at that point, but they also create negative user perceptions of the system and of the acquisition process.

Resource constraints also hamper the DOD test organizations directly. Cuts to budgets and personnel have significantly reduced the number of soldiers, sailors, airmen, and Marines available to serve as users during the test process, especially in the military T&E departments, even as systems have become more complex.[2] This reduced pool of DOD testers impedes the early and close collaboration with systems acquirers and developers that is necessary to support an IID process adequately.

In summary, IT testing in the DOD remains a highly rigid, serial process without the inherent flexibility and collaboration required to support an agile-oriented iterative, incremental development process, particularly as it might be applied to IT systems. If a new IT systems acquisition process is defined and adopted, life-cycle acceptance testing that reflects the IID approach will also be needed in order to achieve success. The rest of this chapter describes how to address the challenges of testing and evalu-

[2] Defense Science Board, *Report of the Defense Science Board Task Force on Developmental Test & Evaluation*, Department of Defense, Washington, D.C., May 2008, available at http://www.acq.osd.mil/dsb/reports/2008-05-DTE.pdf; accessed November 4, 2009.

ation in the DOD acquisition environment in a way that incorporates an IID approach.

"BIG-R" REQUIREMENTS AND "SMALL-R" REQUIREMENTS

For IT systems, a decision point assessment rests on the system's ability to satisfy the stated and approved "big-R" requirements. The term "big-R" requirements in this report refers to a widely recognized purpose, mission, and expected outcome. One example would be a missile system, which would be assessed on the basis of its ability to hit a target at a given range under specified conditions. Another example would be a management information system that would support specific business functional areas and be accompanied by security access levels and a specified data standard and architecture. Such high-level descriptions are expected to be fairly stable over the course of a program, although they may evolve on the basis of feedback from users and other stakeholders.

In contrast, IT systems, particularly software development and commercial off-the-shelf integration (SDCI) IT systems, cannot be expected to have stable requirements initially. The "small-r" requirements referred to in this report are the more detailed requirements, such as those associated with specific user interfaces and utilities, that are expected to evolve within the broader specified architecture as articulated in the initial big-R requirements document. In a sense, small-r requirements could also be thought of as lower-level specifications.

Stated another way, it is challenging if not impossible to accurately capture users' detailed, small-r requirements up front, as their reactions to a prototype or newly fielded system are often negative, even though it may fully meet the specifications set forth in a requirements document. Part of the problem is that there are so many minute details that together contribute to the usability of a system that it is nearly impossible to detail these in advance of giving users an opportunity to try out an actual running system. Usability might be a big-R requirement, but the specific details that would make that happen are small-r requirements that are essentially impossible to specify before users have been able to experiment with the system. Big-R requirements, such as the expected user interface and user paradigms and integration with other concurrently evolving systems and security practices at a high level, will all result in an unpredictable set of specifications and small-r requirements in practice. The need to manage big-R requirements coupled with changing and/or ill-specified small-r requirements is another reason that an iterative, incremental development process is well suited to these types of systems.

INCORPORATING THE VOICE OF THE USER

A significant problem across the DOD in the development and acquisition of IT systems is the lack of ongoing user input, both as a means to determine needed capabilities and as a measure of the success of program development. In the current IT environment, user needs are constantly changing; such constant change is an ever-present factor that breaks any development or testing model which assumes a consistent, comprehensive set of user requirements (both big-R and small-r). Typical DOD IT systems have become so complex that describing either big-R or small-r requirements up front has become severely problematic. In comparison, IID approaches to IT systems rely on user feedback early and at intermediary points to guide development.

Without significant user involvement in developmental testing, the earliest point at which users will be involved may not be until operational testing, far too late in the development process for an IT system. It is crucial to learn from user experiences early in the development process, when a system can be improved more easily and at less expense. If requirements (big-R and especially small-r) are not well understood or are likely to change during the course of the project—conditions that are commonly found in DOD IT developments—an iterative approach with regular and frequent, if not continuous, user feedback will produce results faster and more efficiently than the traditional DOD approach can.

The adoption of IID approaches coupled with a focus on the end-user experience does not mean, however, that other stakeholders and nonfunctional requirements (such as information assurance, reliability, and so on) are unimportant. Historically, other stakeholder voices have dominated the process to the exclusion of the end user. The committee urges a rebalancing and a focus on end-user mission success that incorporates higher-level architectural and nonfunctional requirements at appropriate phases of system development and deployment.

TOWARD CONTINUOUS OPERATIONAL ASSESSMENT

Agile approaches and iterative, incremental approaches to software development have been receiving increased attention in industry in recent years and are having a significant impact on how software is developed and how systems are tested. While architectural and systems-level engineering considerations will continue to be significant drivers of system success, the shift toward IID and agile processes also appropriately requires the incorporation of user perspectives throughout the development life cycle. In the case of assembling commercial off-the-shelf (COTS) solutions, the criteria for solutions that please users are often related less to the technical architecture that accounts for how the components are put together than to the workflow of the resulting system—How easy is it to perform specific tasks with the system? How easy is it to maintain, manage, or perform upgrades on the system? Answers to these questions cannot typically be answered up front but rather are best answered through end-user experiences. In the case of any new systems development, user input is critical to the requirements and deployment-readiness of the resulting system, but generally users are unable to specify the details of these needs in the requirements phase. Thus, requirements in IT systems are best guided through user input based on direct experience with current prototypes. The acquisition approach described in this report incorporates the restructuring of the DOD testing and acceptance process to be the enabler of this user input, and the incorporating of operationally realistic testing with user feedback into a routine of continuous operational assessment.

A continuous operational assessment process would replace the traditional DOD development, testing, and acquisition process. A continuous process would need to encompass two critical features: an acceptance team and a metrics collection and reporting capability:

- *An acceptance team (AT)* would be charged with providing feedback on the acceptability of the solution toward meeting the users' goals. The AT should be drawn from the expected initial users of the IT system and ideally should be the cadre that first tests the system in the field and then trains subsequent users in use of the system. The AT would work with the development team and acquisition (program management) team through the entire process. The AT would initially engage with these teams as high-level requirements are gathered and initial big-R requirements documents are published, and then it would continue in this relationship through to product deployment. An important function of the AT would be to keep the requirements, functional deliverables, and test plans synchronized.
- A robust *metrics collection and reporting capability (MCRC)* would collect, aggregate, and analyze end-user service consumption and experiences. The MCRC—leveraging existing operational tools, including enterprise monitoring and management capabilities, DOD information assurance capabilities, and commercial capabilities in combination with measured end-user performance—would provide visibility into what functions end users are actually using in accomplishing their day-to-day missions and how they are using them. The MCRC would provide all stakeholders with a clear indication of operational effectiveness based on actual operational use.

Acceptance Teams

The development process continues through a set of iterations as described earlier in the report. In each iteration, the AT would fill the role of "the customer" on matters pertaining to the achievement of iteration objectives. The AT would evaluate each iteration prototype and identify issues and plausible changes to satisfy user acceptance of stated requirements before the development process went forward. Thus, the development team would secure early feedback from the AT on unclear, misunderstood, or incorrect requirements. For each iteration, the AT, in its role as proxy for the ultimate customer, would validate the current requirements list while the development team would provide work estimates and cost projections and develop a plan for the next iteration.

A responsibility of the AT in assisting the development process would include building (or working with the development team to build) acceptance tests for each iteration. This approach is a cornerstone of *Test-Driven Development: By Example,*[3] in which tests for a deliverable are written first, and then a system to satisfy those tests is developed. Using tests in this manner is particularly helpful with larger, more complex organizations as it helps remove ambiguity or gaps in communication.

Moving to an IID methodology does not imply any reduction in oversight or the shepherding of acquisition funds. To ensure satisfactory progress of programs, periodic and regular progress checkpoints need to occur with the acquisition executive. In keeping with agile-inspired methodologies, these checkpoints should be based on calendar or funding milestones rather than being requirements-driven progress points. In agile processes, iterations are based on time-boxing work schedules, whereby some content may slip from one iteration to the next, but iterations are closed according to the schedule, thus allowing for prompt identification of erroneous estimates of the time required to complete work items and ensuring continuous user input regarding priorities. In a similar manner, checkpoints with acquisition executives could be based either on time duration or on a funding milestone (as the two are frequently closely correlated in an IT project). These checkpoints should be less frequent than one per iteration, perhaps occurring every 6 to 18 months.

In these checkpoints, the AT would be the key voice in articulating the "value delivered." It might be that at a checkpoint, the requirements delivered are not as anticipated at the start of the program, but the value to the end users is still quite significant. In such cases, the acquisition executive would examine and understand the reasons for the deviation in

[3] Kent Beck, *Test-Driven Development: By Example,* Addison-Wesley, Old Tappen, N.J., 2003.

the plan and take into account user reactions to progress in the program as far as it had gone. Conversely, a program might be tracking exactly to the initial schedule but could be producing an asset with which the AT is not satisfied. In such cases the acquisition executive should intervene and determine if corrective action is possible or if the fault lies in the concept of the program itself. In either case, the acquisition executive is getting regular feedback on project progress with a clear opinion from those who will benefit from the system.

OT&E is, at its core, about determining the effectiveness and suitability of a system for deployment. During development and before a system reached OT&E, the AT would carry this responsibility. The AT would recommend to the acquisition executive when an iteration of an IT system was ready for deployment. Such deployments can take several forms. Initial deployments may be limited in scope, with the intention of testing system effectiveness or of allowing some features to be exploited while others continue to be developed (e.g., joint programs that are initially deployed with a single Service). Other deployments may be wider in scope to allow value to be captured from these systems while more advanced features are developed in future iterations. An important difference in this process is that deployment is not a one-time event at the end of the program. The AT would recommend deployment and the scope of deployment on the basis of the operational benefit, functional value to the user, and risk of deployment, but would not need to (and should not) wait until "completion" of the entire system before making such recommendations.

Evaluation Through Operational Use Metrics

One of the major lessons learned from interactions with commercial suppliers of IT systems is that significant benefits come from these suppliers' understanding of and reliance on actual end-user behavior as measured by actual end-user actions. In fact, the instrumentation of products and services for the collection of actual end-user metrics is so engrained in many commercial IT companies as to be unremarkable to commercial suppliers. Indeed, because this practice is second nature to those in the commercial sector, it was necessary for the committee to expressly solicit comments on this point from briefers. Commercial IT companies drive their entire investment portfolios on the basis of anticipated and actual end-user consumption patterns and/or end users' engagement with their offered products and services. Those products and services with large and committed user bases drive a preponderance of the businesses' valuation and receive commensurate corporate leadership attention and investment. Small changes in interaction patterns are induced and then measured and analyzed minutely for an understanding of how best to improve the user experience and increase user satisfaction as measured by user engagement. As a result, virtually every aspect of the business is focused on satisfying end-user needs as quickly as possible, with the associated increase in network-based productivity that has been witnessed on the World Wide Web.

Many tools can be used to gather this information, including run-time configuration management, run-time collection and reporting, near-real-time aggregation, and business analytics. Many of the supporting tools are also used by service operations to identify early signs of technology outages or slowdowns and to begin assessing and diagnosing a problem before it becomes a catastrophic failure.

This report recommends incorporating the voice of the end user at all stages of the system life cycle. From a test perspective, such incorporation focuses resources in a number of important ways:

- Those services that are integral to every higher-level capability receive special attention during testing and are added incrementally much more deliberately, which may impact fielding cycle time;
- Those services that are exercised most strenuously and frequently by end users directly experience extensive, highly iterative beta testing with actual successful end-user sessions forming the basis for a determination of operational effectiveness; and
- Field failures are automatically reported and incorporated into development and testing procedures as necessary, resulting in "living" test documents.

Establishing an MCRC along with leveraging current DOD tools and available commercial tools and practices would overtly move the operational evaluation assessment from a speculative proposition based on surrogate run times, users, test data, and marginally current requirements specifications, to a managed and measured investment assessment based on current, actual end-user missions and needs.

INCORPORATING COMMON SERVICES DEFINITIONS

For years, IT systems developers have employed functionalities that are externally supplied and operationally validated as a basis for their success, without expecting to revalidate those functionalities as part of their formal test regimen. Examples include the use of the Internet Protocol (IP) and the Domain Name System (DNS). These capabilities are provided externally to the capability being tested, have already been validated in separate acquisitions, and thus need not be included in the scope of the IT systems test regimen. As more services have been commoditized and/or supplied through commercial means, IT systems acquisition practice has not changed to address this reality. Unfortunately, no mechanisms exist to identify and track these supplied services or to apply a consistent approach for their use throughout the current DOD acquisition process. This is another negative repercussion of the weapons systems-based acquisition approach, where far fewer opportunities for shared services exist.

For example, the DOD has broadly adopted a set of networking capabilities that are integral to every IT system and that do not require revalidation (e.g., the IP and DNS capabilities mentioned above). As more of the technology stack becomes commoditized or provided as a service, the set of associated capabilities not requiring revalidation should likewise grow. In addition, testing approaches should be broadened to account for this commoditization. Developers of common service capabilities should account for the full range of possible application, from strategic, fixed-base IT systems to tactical, end-user-focused IT systems. Similarly, the testing of common services should reflect the full intended scope of application of the services. Dependent developers should be permitted and encouraged to view these common services as operationally validated externally, as long as they adhere to the terms of the supplied service. These developers' test teams should accept and treat these as externally supplied and acceptable operational services, regardless of the validating organization.

Commercial off-the-shelf hardware, software, and services (CHSS) IT systems acquisition can similarly better leverage commercial experience in place of formal DOD testing and oversight review. Service-level agreements established for commercial services—which often have been validated by thousands or millions of users or hundreds of commercial entities—constitute, in effect, a test environment whose results should be accepted prima facie. Past validation of platform components—either validation from widespread use in the commercial marketplace or prior validation in an SDCI IT system—can also substitute for new testing.

As an example, consider a public key infrastructure (PKI) that is sourced and validated as a well-defined service on which secure identity management will depend. As a result, the PKI program manager must architect the supplied service to support the range of users anticipated. The PKI test regimen should address the specified terms of service that result for the PKI. So long as a dependent PM uses the standard service interface and has no requirements that exceed the already-validated PKI service scope, the PKI should be treated as an existing and validated external service that is outside the scope of the dependent PM's formal testing regimen. Other examples include "on-demand" computing and storage, network-centric enterprise services, and visualization services.

This approach to the evaluation of common services avoids the cost and time required to recertify proven products individually. Risks of undetected failures in these products are mitigated by development tests of integrated modules and operational testing when the composite system undergoes rigorous evaluations to determine effectiveness and suitability.

VIRTUAL INFORMATION TECHNOLOGY TEST ENVIRONMENTS

The use of integrated virtual information technology test environments may be one way to facilitate testing that would allow early prototypes of systems to be subjected to much more realistic test conditions, thereby helping to identify potential problems in development as soon as possible. Such test environments would rely on a distributed test network that could be accessed by both government and industry, when appropriate, for use in performing early acceptance testing. A broad range of simulation systems and operational command-and-control systems that can represent realistic operational elements would provide the necessary data to drive such systems during testing. Linking the proponents of these simulations and systems through a distributed network would allow them to maintain the systems within their existing facilities while also providing opportunities for use during larger test events. Additional applications necessary to control, monitor, and log data during such tests would augment the sets of simulations and systems.

It is important that virtual IT test environments have the ability not only to test the basic functionality of systems but also to emulate as much of the expected operational environment as possible. One of the recommendations of the National Research Council report *Testing of Defense Systems in an Evolutionary Acquisition Environment* was to "revise DOD testing procedures to explicitly require that developmental tests have an operational perspective (i.e., are representative of real-world usage conditions) in order to increase the likelihood of early identification of operational failure modes."[4] A simulation-based test environment has the potential to provide such functionality, as has been shown in multiple DOD training and experimentation environments.

The technology necessary to achieve virtual test environments is already well established. In fact, multiple (somewhat duplicative and overlapping) programs that have similar capabilities for doing exactly this kind of testing already exist within the DOD. These programs may

[4] National Research Council, *Testing of Defense Systems in an Evolutionary Acquisition Environment*, The National Academies Press, Washington, D.C., 2006.

provide an important starting point from which an expanded capability for early and continuous acceptance testing could be implemented. A sampling of such programs from across the military Services and defense agencies includes the following:

- The *Systems of Systems Integration Laboratory (SoSIL)* is a large-scale communications network for modeling and simulation, hardware and software integration, and virtual operational testing; SoSIL also offers a "soldier-in-the-loop" capability.[5]
- The Army's *Cross Command Collaboration Effort* is an effort to establish and evolve a consistent and core set of modeling and simulation tools, data, and business processes that meet the common environment requirements of the U.S. Army's Training and Doctrine Command, Army Test and Evaluation Command, and Research, Development and Engineering Command. This common environment will facilitate those organizations' interoperability with the materiel development community to help conduct the distributed development of doctrine, organizations, training, materiel, leadership and education, personnel, and facilities.[6]
- The *Air Force Integrated Collaborative Environment* (AF ICE) is intended to provide a persistent, composable, flexible infrastructure along with a series of tools, standards, processes, and policies for using the environment to conduct the continuous analysis required to support a capabilities-based planning process.[7]
- The *Joint Mission Environment Test Capability* (JMETC) was established in October 2006 to "link distributed facilities on a persistent network, thus enabling customers to develop and test warfighting capabilities in a realistic joint context."[8] JMETC has already established a persistent test network, through the Secret Defense Research and Engineering Network, which provides connectivity to both Service and industry assets. It relies on the Test and Training Enabling Architecture (TENA) as its infrastructure for data exchange; TENA provides a standard object model and interfaces to the Distributed Interactive Simulation Protocol and the

---

[5] Boeing, *FCS Systems of Systems Integration Laboratory Backgrounder,* May 2007, available at www.boeing.com/defense-space/ic/fcs/bia/080523_sosil_bkgndr.pdf; accessed November 12, 2009.

[6] Brian Hobson and Donald Kroening, "Cross Command Collaboration Effort (3CE)," *Spring Simulation Multiconference: Proceedings of the 2007 Spring Simulation Multiconference,* Vol. 3, 2007.

[7] B. Eileen Bjorkman and Timothy Menke, "Air Force-Integrated Collaborative Environment (AF-ICE) Development Philosophy," *ITEA Journal of Test and Evaluation* 27(1), March/April 2006.

[8] Richard Lockhard and Chip Ferguson,"Joint Mission Environment Test Capability (JMETC)," *ITEA Journal* 29:160-166, 2008.

High Level Architecture, which are widely used standards for modeling and simulation. JMETC has already established linkages to the Future Combat System program and AF ICE.[9]

- The *Distributed Engineering Plant* (DEP) was established in 1998 by the Naval Sea Systems Command (NAVSEA) to identify and resolve combat battle management command, control, communications, and computers (C4) systems interoperability problems prior to deploying new and upgraded systems to sea. Enabled by today's newest networking technology, DEP links the Navy's shore-based combat systems/C4/hardware test sites, which are located in geographically disparate facilities across the nation, into a virtual shore-based battle group that exactly replicates a battle group fighting at sea. By inserting "ground truth" system simulation and stimulation data and then observing how the combat systems exchange and display tactical data, engineers can identify precisely and solve interoperability problems ashore well before those systems enter the operating forces. This approach emphasizes shore-based testing and warfare systems integration and interoperability testing and acceptance certification of operational IT systems in a test environment similar to their ultimate shipboard operational environment; it also emphasizes interoperability assessments, which are a prerequisite for the operational certification of the ships in strike force configurations prior to deployment.

Obviously, numerous organizations across the DOD with roles and missions oriented toward testing and evaluation may also have capabilities that could be leveraged for such an effort. Among them are the Joint Interoperability Test Command, the Army Test and Evaluation Command, the Air Force Operational Test and Evaluation Center, and the Navy Operational Testing and Evaluation Force. Numerous software testing and distributed testing capabilities also exist in organizations such as the Defense Advanced Research Projects Agency and the various research laboratories within the Services.

Establishing the technical underpinnings of virtual IT test environments is only part of the solution to improving early testing in acquisition. In addition, appropriate policy and process changes would need to be implemented to mandate activities that would utilize such an environment. Some of the issues that must be addressed include data sharing across a testing enterprise, the establishment of standards for data exchange, and the formalizing of the role of early testing in acquisi-

[9] Test and Training Enabling Architecture Software Development Activity, *JMETC VPN Fact Sheet*, Central Test & Evaluation Investment Program, U.S. Joint Forces Command, Department of Defense, Washington, D.C., November 23, 2009, available at https://www.tena-sda.org/download/attachments/6750/JMETC-VPN-2009-11-23.pdf?version=1.pdf.

tion—likely requiring revisions to DODI 5000.2. Other issues include the governance and management of such a capability and the roles and responsibilities in terms of executing testing in such an environment. Some of these issues are raised in the DOD's *Acquisition Modeling and Simulation Master Plan*, issued in April 2006, which focuses on improving the role of modeling and simulation for testing.[10]

Another challenge in making such environments usable is to ensure that the complexity required to perform the integration of systems and configuration for tests is minimized; otherwise, the costs of using such test environments would far outweigh the benefits. Paramount in managing the complexity involved is the establishment of a formal systems engineering process involving test design, integration, documentation, configuration management, execution, data collection, and analysis. Also important is the establishment of standards for simulation and systems interoperability that allow for common interfaces and the reuse of systems that also provide enough flexibility to adapt to new requirements. And finally, a management process must be married to the systems engineering process so that users are invited to participate and incentives are created for industry and government to share data and work toward common goals.

[10] Department of Defense, *Acquisition Modeling and Simulation Master Plan*, Software Engineering Forum, Office of the Under Secretary of Defense (Acquisition, Technology and Logistics) Defense Systems, Washington, D.C., April 17, 2006.

Bibliography

In addition to the specific sources cited in this report's footnotes and information provided in briefings to the committee (the briefings are listed in Appendix E), the committee also found the following list of resources useful.

Association for Enterprise Integration (AFEI). 2008. *Industry Recommendations for DoD Acquisition of Information Services and SOA Systems.* SOA AcquisitionWorking Group, AFEI Executive Forum on Business Change, Arlington, Va., July 7.

Beebe, H., and D. Meyerriecks. 2008. *Next Generation C2: Formalizing Military Mashups.* Prepared for the North Atlantic Treaty Organization (NATO-OTAN) by representatives of the Defense Information Systems Agency, Arlington, Va.

Boudreau, M. 2006. *Acoustic Rapid COTS Insertion: A Case Study in Spiral Development.* Acquisition Research Case Study. Acquisition Research Program, Graduate School of Business and Public Policy, Naval Postgraduate School, Monterey, Calif., October 30.

Carstens, R.D., M.A. Cohen, and M.F. Kupcy. 2008. *Changing the Culture of Pentagon Contracting: A Publication of the Privatization of Foreign Policy Initiative.* New America Foundation, Washington, D.C.

Clancy, M., T. Goins, E. Giorgio, J. Gosler, J. Levine, D. Meyerriecks, R. Perlman, M. Perlman, D. Souleles, and B. Tribble. 2008. "NITT IAG Observations and Recommendations." PowerPoint presentation, U.S. Department of Defense, Defense Information Systems Agency, June 6.

Dahmann, J.S. 2008. "Changes to DoD Acquisition Process on the Way." September 12.

Defense Acquisition Performance Assessment Panel. 2006. *Defense Acquisition Performance Assessment Report.* U.S. Department of Defense, Washington, D.C., January. Available at http://www.afei.org/documents/DAPA-Report-web-feb21.pdf; accessed July 15, 2008.

Defense Acquisition University. 2006. "Defense Acquisition Performance Assessment (DAPA) Project: List of Documents." Defense Acquisition University, David D. Acker Library and Knowledge Repository, Fort Belvoir, Va., March. Available at http://www.dau.mil/library/pdf/DAPA.pdf; accessed July 15, 2008.

Defense Acquisition University. 2008. "DoD Announces Major Revision to Acquisition Policy." Available at https://akss.dau.mil; accessed January 27, 2009.

Deputy Under Secretary of Defense, Advanced Systems and Concepts. 2006. *Joint Capability Technology Demonstration (JCTD): Program Overview.* U.S. Department of Defense, October.

DISA (Defense Information Systems Agency). 2008. *A Guide for DISA Small Project Execution.* U.S. Department of Defense, DISA, Office of the Component Acquisition Executive, Washington, D.C., June 30.

DISA. 2008. *Cooperative Review.* U.S. Department of Defense, DISA, Washington, D.C., August.

DISA. 2008. *Enabling Warfighting: Speed, Services and Capabilities.* U.S. Department of Defense, DISA, Washington, D.C. Available at http://www.gcn.com/acmfiles/sponsored_papers/DISA-final.pdf; accessed July 18, 2008.

England, G. 2009. "Investment Review Board (IRB) Roles and Responsibilities." Deputy Secretary of Defense, Directive-Type Memorandum (DTM) 08-020, OSD 15508-08, January 26.

Executive Office of the President. 1986. *A Formula for Action: A Report to the President on Defense Acquisition by the President's Blue Ribbon Commission on Defense Management*. April. Available at http://www.ndu.edu/library/pbrc/36ac7c2.pdf; accessed July 15, 2008.

GAO (Government Accountability Office). 2008. *Defense Acquisitions: A Knowledge-Based Funding Approach Could Improve Major Weapon System Program Outcomes.* GAO/08-619. Washington, D.C., July. Available at http://www.gao.gov/new.items/d08619.pdf; accessed July 15, 2008.

GAO. 2008. *DoD Business Systems Modernization: Key Marine Corps System Acquisition Needs to Be Better Justified, Defined, and Managed.* GAO-08-822. Washington, D.C., July. Available at http://www.gao.gov/new.items/d08822.pdf; accessed August 6, 2008.

Gates, Robert M. 2009. Statement Submitted by Secretary of Defense Robert M. Gates to the House Armed Services Committee, U.S. Congress, January 27.

Grasso, A. 2009. *Information Technology Acquisition: A Common Sense Proposal.* Defense AT&L, March-April.

Kerber, R. 2007. "Raytheon: Army Aware of Missile's Flaw, Firm Being Sued in Pilot's Death." *Boston Globe,* December 26. Available at http://www.boston.com/business/technology/articles/2007/12/26/raytheon_army_aware_of_missiles_flaw/; accessed July 15, 2008.

National Research Council. 1999. *Realizing the Potential of C4I: Fundamental Challenges.* National Academy Press, Washington, D.C.

Office of the Under Secretary of Defense for Acquisition, Technology, and Logistics. 2007. *Report of the Defense Science Board Task Force on Developmental Test and Evaluation.* U.S. Department of Defense, Washington, D.C., September. Available at http://www.acq.osd.mil/dsb/reports/2008-05-DTE.pdf; accessed July 15, 2008.

Office of the Under Secretary of Defense for Acquisition, Technology, and Logistics. 2007. *Report of the Defense Science Board Task Force on Mission Impact of Foreign Influence on DoD Software.* U.S. Department of Defense, Washington, D.C., September. Available at http://www.acq.osd.mil/dsb/reports/2007-09-Mission_Impact_of_Foreign_Influence_on_DoD_Software.pdf; accessed July 15, 2008.

Office of the Under Secretary of Defense for Acquisition, Technology, and Logistics. 2008. "Memorandum for Chairman, Defense Science Board: Terms of Reference—Defense Science Board Task Force on the Department of Defense Policies and Procedures for the Acquisition of Information Technology." U.S. Department of Defense, Washington D.C.

Stavros, N., M. Dettman, and J. Albrant. 2007. *Engineering Governance. Space and Naval Warfare Systems Center, San Diego.*

Thibodeau, P. 2008. "Pentagon's IT Unit Seeks to Adopt Cloud Computing." *New York Times,* July 17. Available at http://www.nytimes.com/idg/IDG_852573C400693880002574890080F9EF.html?partner=rssnyt&emc=rss; accessed July 18, 2008.

Vanucci, S. 2008. "New Acquisition Policy and Its Impact on Systems Engineering." Systems and Software Engineering/Enterprise Development, Office of the Deputy Under Secretary of Defense for Acquisition, Technology and Logistics. Presentation at the NDIA 11th Annual Systems Engineering Conference, October 21.

Appendixes

Appendix A

Brief Overview of the Defense Acquisition System for Information Technology

THE DEFENSE ACQUISITION MANAGEMENT SYSTEM

The Defense Acquisition Management System (DAMS), defined in DOD Instruction 5000.2, specifies a single framework to address both information technology (IT) systems (termed "automated information systems" [AISs] in DOD regulations) and weapon systems. The milestone decision process defined in the instruction and applied to the acquisition of both weapon systems and information technology is depicted in Figure A.1.

DODI 5000.2 allows for differentiation of the prescribed acquisition process based on the underlying technological maturity. It identifies evolutionary acquisition as the "preferred strategy for the rapid acquisition of mature technology" and notes that in an evolutionary acquisition strategy, close cooperation is required between users, testers, and developers. The DOD 5000.2 milestone decision process flow for evolutionary acquisition is shown in Figure A.2. It further states that "MDAs [milestone decision authorities] may tailor regulatory program information to fit the particular conditions of an individual program." Implicit in this statement is that regulatory program information must still be provided, since the MDAs may tailor only the instruction implementation.

Joint Capabilities Integration and Development System

The Joint Capabilities Integration and Development System (JCIDS) supports the Chairman of the Joint Chiefs of Staff and the Joint Require-

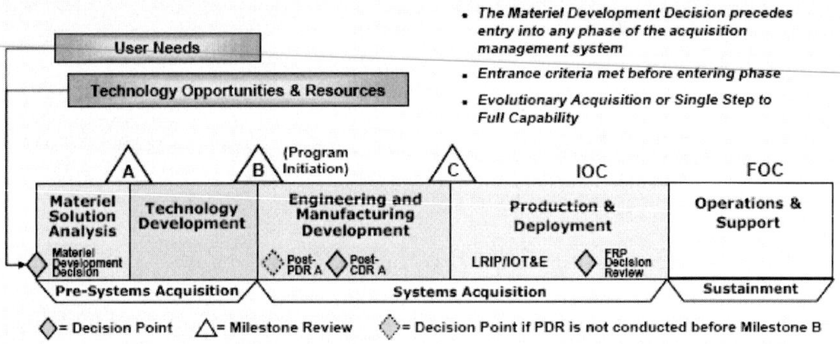

FIGURE A.1 The milestone decision governance and oversight process of the Defense Acquisition Management System applied to both weapon systems and automated information systems.

ments Oversight Council (JROC) in identifying, assessing, and prioritizing joint military capability needs as required by law. The capabilities are identified by analyzing what is required across all functional areas to accomplish the mission.

The JROC recognizes that the same level of oversight is not required for all information systems. Therefore, information systems are divided into four categories with appropriate oversight for each:

- Information systems with a post-Milestone B developmental cost of less than $15 million are not subject to joint oversight or approval under the JCIDS process. The sponsor manages the requirements, approves the JCIDS documents, and complies with appropriate acquisition requirements.
- Information systems that are defense business systems, regardless of cost, are to comply with the process defined by the Defense Business Systems Management Committee. These systems will employ a business case document using the business capability life-cycle process in lieu of an Initial Capabilities Document/Capabilities Development Document (ICD/CDD) to justify the need for a solution. In those cases where the JCIDS gatekeeper, on the advice of the lead functional capabilities board (FCB), determines that joint oversight of the business system is required, the business case document will be reviewed and validated in lieu of the appropriate JCIDS documents.
- Information systems that are an integral part of a weapon or weapon system and enable weapon capabilities are considered to be part

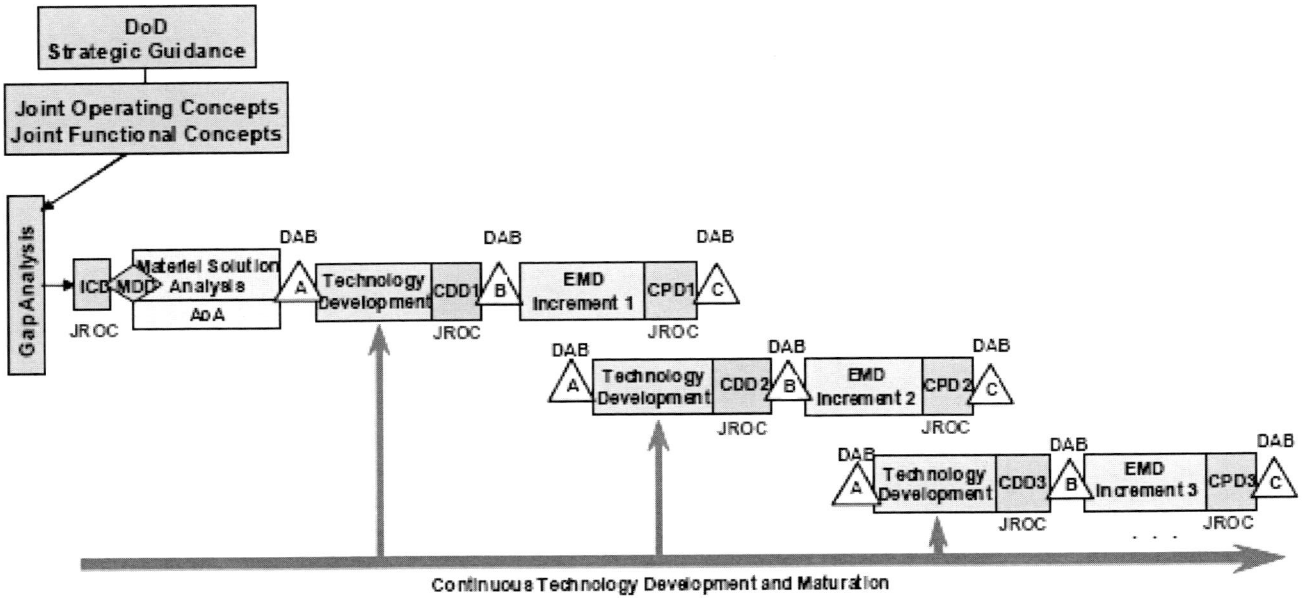

FIGURE A.2 The milestone decision governance and oversight process of the Defense Acquisition System adapted for evolutionary acquisition.

of the weapon system program and do not require separate JCIDS documents or oversight.

- Information systems that provide capabilities through software development and integration with commercial off-the-shelf hardware require an ICD for initiation of new-capability development. The CDD will support the development and fielding process. A CPD is not required unless the program is going through a formal Milestone C decision and the MDA requires it.

The Joint Staff Director of Force Structure, Resources, and Assessment/Requirements Management Division (J-8/RMD) and/or the Lead FCB will make a determination if it is not clear which definition applies to a particular information system.[1]

The JCIDS processes, as illustrated in Figure A.3, overlay and support the Defense Acquisition Management System. The JROC decision tree and membership are illustrated in Figure A.4.

Recently, the JCIDS process for information technology systems was changed to reflect the evolving nature of requirements for IT systems. This new JCIDS policy[2] recognizes the need for the JROC to focus on top-level requirements at the beginning of a program and delegates refinement of subsequent requirements to a lower-level flag-officer-level body.

Planning, Programming, Budgeting, and Execution System

The purpose of the PPBES is to allocate resources within the DOD. It is important for program managers and their staffs to be aware of the nature and timing of each of the events in the PPBE process, since they may be called on to provide critical information that could be important to program funding and success.

In the PPBE process, the Secretary of Defense establishes policies, strategy, and prioritized goals for the DOE that are subsequently used to guide resource allocation decisions that balance the guidance with fiscal constraints. The PPBE process consists of four distinct but overlapping phases:

- *Planning*. The planning phase of PPBE, which is a collaborative effort by the Office of the Secretary of Defense and the Joint Staff, begins with a resource-informed articulation of national defense policies and military strategy known as the Strategic Planning Guidance.

---

[1] CJCSI 3170.01G, Enclosure B, 2009.
[2] Laura Knight, Net-Enable Command Capability briefing, DISA, October 28, 2008.

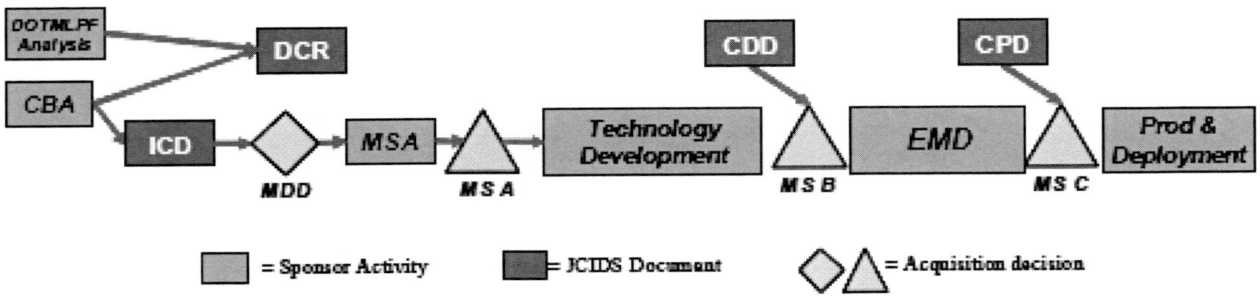

FIGURE A.3 JCIDS process and acquisition decisions. SOURCE: Chairman of the Joint Chiefs of Staff Instruction 3170.01G, Enclosure A, March 2009.

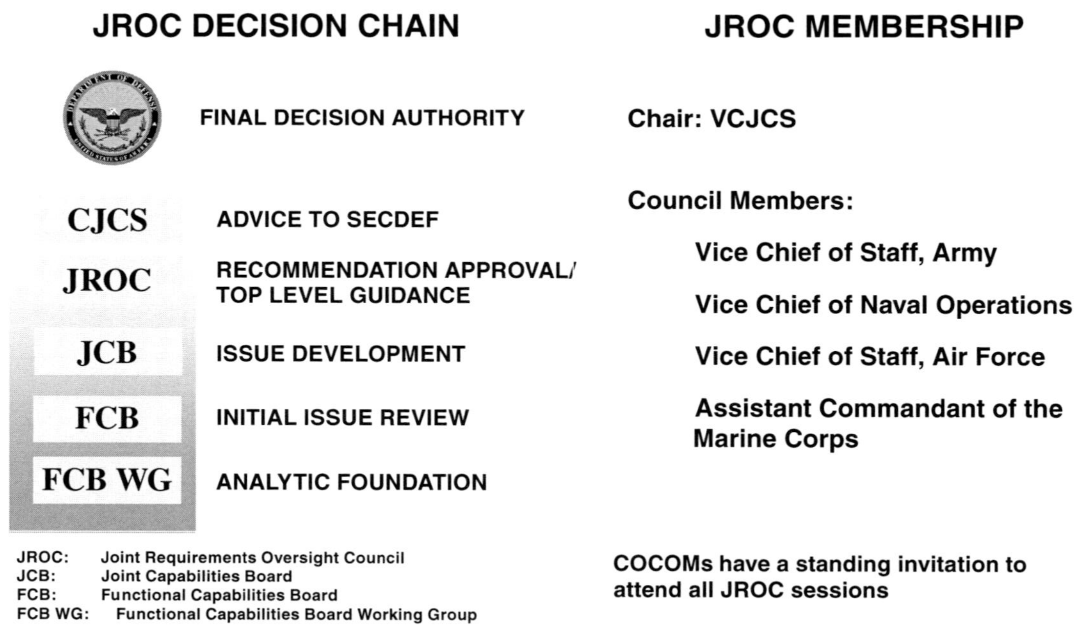

FIGURE A.4 JROC decision tree and membership. NOTE: Functional capability boards (FCBs) exist for command and control, battlespace awareness, logistics, force support, force protection, force application, and network-centric operations, focused logistics, force management, and joint training. FCBs are led by general-officer-level JCS staff. Membership consists of O6-level representation of the Services, the Combatant Commands, OSD(AT&)L, OSD(I), ASD(NII)/DOD CIO, Director of PA&E, DIA, and the specific FCB Executive Secretary. SOURCE: Pat Wills, JCIDS Brief, Defense Acquisition University, January 2009, available at https://acc.dau.mil, accessed June 2009.

- *Programming.* The programming phase begins with the development of a program objective memorandum (POM) by each DOD component.
- *Budgeting.* The budgeting phase of PPBE occurs concurrently with the programming phase; each DOD component submits its proposed budget estimate simultaneously with its POM.
- *Execution.* The execution review occurs simultaneously with the program and budget reviews. The purpose of the execution review is to provide feedback to the senior leadership concerning the effectiveness of current and prior resource allocations.[3]

The PPBES, as currently implemented, follows a biennial cycle to reduce the number and amount of budget artifacts provided and reviewed by DOD components and OSD/JCS, respectively, on an annual basis. New initiatives, theoretically, can be started in any budget year; however, the activities required in an "off" year are based on exception processing, as opposed to the normal budget process. However, even during an "on" year, the window to successfully present and argue for a new initiative or a major change in an initiative is formally from mid-August to mid-October.

INFORMATION TECHNOLOGY AND THE DEFENSE ACQUISITION MANAGEMENT SYSTEM

Although some specific requirements of DOD acquisition regulations apply only to weapon system programs or to AIS programs, the same overall program structure template and milestone decision process are applied to both. A key facet of the DAMS is a series of thresholds that establish the acquisition category (ACAT) of an acquisition program, and with that determination also establish the MDA responsible for oversight of the acquisition program. The largest or most highly visible programs are designated as major defense acquisition programs (MDAPs) or major automated information system (MAIS) programs, and the MDA for them is assigned to the Defense Acquisition Executive (DAE), Under Secretary of Defense for Acquisition, Technology and Logistics (USD AT&L), or to the Service or component acquisition executives within the military Services or defense agencies.

The criterion that governs the assignment to specific acquisition categories is codified in Title 10 U.S.C. Chapter 144 and 144A and is provided in DOD 5000[4] as shown in Table A.1. The expenditure-level

[3] Defense Acquisition University (DAU), *Defense Acquisition Guidebook*, DAU, Department of Defense, Washington, D.C., 2009.

[4] DODI 5000.2, 2008.

TABLE A.1 Criteria for Acquisition Category Designation

Acquisition Category	Reason for ACAT Designation	Decision Authority
ACAT I	• MDAP (section 2430 of Reference (k)) ○ Dollar value: estimated by the USD(AT&L) to require an eventual total expenditure for research, development, test and evaluation (RDT&E) of more than $365 million in fiscal year (FY) 2000 constant dollars or, for procurement, of more than $2.190 billion in FY 2000 constant dollars ○ MDA designation • MDA designation as special interest	ACAT ID: USD(AT&L) ACAT IC: Head of the DOD Component or, if delegated, the CAE (not further delegable)
ACAT IA[1,2]	• MAIS (Chapter 144A of title 10 of U.S.C. (Reference (k)): A DOD acquisition program for an Automated Information System[3] (either as a product or a service) that is either: ○ Designated by the MDA as a MAIS; or ○ Estimated to exceed: — $32 million in FY 2000 constant dollars for all expenditures, for all increments, regardless of the appropriation or fund source, directly related to the AIS definition, design, development, and deployment, and incurred in any single fiscal year; or — $126 million in FY 2000 constant dollars for all expenditures, for all increments, regardless of the appropriation or fund source, directly related to the AIS definition, design, development, and deployment, and incurred from the beginning of the Materiel Solution Analysis Phase through deployment at all sites; or — $378 million in FY 2000 constant dollars for all expenditures, for all increments, regardless of the appropriation or fund source, directly related to the AIS definition, design, development, deployment, operations and maintenance, and incurred from the beginning of the Materiel Solution Analysis Phase through sustainment for the estimated useful life of the system. • MDA designation as special interest	ACAT IAM: USD(AT&L) or designee ACAT IAC: Head of the DOD Component or, if delegated, the CAE (not further delegable)

ACAT II	• Does not meet criteria for ACAT I • Major system ○ Dollar value: estimated by the DOD Component Head to require an eventual total expenditure for RDT&E of more than $140 million in FY 2000 constant dollars, or for procurement of more than $660 million in FY 2000 constant dollars (section 2302d of Reference (k)) ○ MDA designation[4] (paragraph (5) of section 2302 of Reference (k))	CAE or the individual designated by the CAE[4]
ACAT III	• Does not meet criteria for ACAT II or above • AIS that is not a MAIS	Designated by the CAE[4]

[1]In some cases, an ACAT IA program, as defined above, also meets the definition of an MDAP. The USD(AT&L) shall be the MDA for such programs unless delegated to a DOD Component. The statutory requirements that apply to MDAPs and MAIS shall apply to such programs.

[2]The MDA (either the USD(AT&L) or, if delegated, the ASD(NII)/DOD CIO or another designee) shall designate MAIS programs as ACAT IAM or ACAT IAC. MAIS programs shall not be designated as ACAT II.

[3]Automated Information System: A system of computer hardware, computer software, data, or telecommunications that performs tasks such as collecting, processing, storing, transmitting, and displaying information. Excluded are computer resources, both hardware and software, that are:

a. an integral part of a weapon or weapon system;

b. used for highly sensitive classified programs (as determined by the Secretary of Defense);

c. used for other highly sensitive information technology programs (as determined by ASD(NII)/DOD CIO); or

d. determined by the USD(AT&L) or designee to be better overseen as a non-AIS program (e.g., a program with a low ratio of RDT&E funding to total program acquisition costs or that requires significant hardware development).

[4]As delegated by the Secretary of Defense or Secretary of the Military Department.

SOURCE: DODI 5000.2.

thresholds for MAIS programs are significantly lower than they are for MDAP programs, subjecting more AIS programs to increased oversight and governance from the highest levels of the DOD. In fact, the overall program size thresholds for the ACAT IAM designation (where the M denotes a MAIS ACAT I program) for AIS programs are smaller than the corresponding ACAT II for non-AIS programs, and there is no ACAT II for AIS programs.

As a consequence of these differences in program threshold levels for oversight between MAIS programs and MDAP programs, the former are subjected to the same level of intensive management as are the ACAT 1D weapon systems programs. This level of management has resulted in the ponderous oversight of ACAT IAM IT programs that are funded at much lower levels. Specifically, an IT program funded at $32 million for research, development, test, and evaluation (RDTE) in any fiscal year gets the same level of oversight as a weapon system program funded at $365 million or more over its RDTE phase. A more extreme metric is that an IT program funded with a total life-cycle cost (including operation and maintenance) of $378 million receives the same level of oversight as a weapon system program funded at $2.190 billion in procurement funding. Moreover, the two-tier acquisition category definitions for IT programs appear to drive more systems into the "major" category (and thus oversight by the Office of the Secretary of Defense) than do the weapon system acquisition category definitions, which are three-tiered.

The governance structure that ultimately brings recommendations forward to the MDA for MDAP or MAIS programs varies as a function of the types of programs, but in virtually every case it is a four-tier process with the program management office (PMO) at the bottom responsible for preparing for a milestone decision review; a collection of integrated product teams (IPTs) with an overarching integrated product team (OIPT) to work with the PMO and provide review and oversight in preparation for a milestone decision review; a formal decision body with a group such as the Defense Acquisition Board (DAB) or Information Technology Acquisition Board (ITAB) to advise the MDA; and the MDA as the responsible party in making the milestone decision at each of the key Milestone A, B, and C decision points identified in Figures A.1 and A.2. This structure is depicted for various types of programs in Figure A.5.

The formal decision forums are key bodies in this oversight and governance process. It is instructive to examine their composition. DAB board members and advisors are listed in Table A.2.[5] Although some specific

[5] Defense Acquisition University (DAU), *Defense Acquisition Guidebook*, DAU, Department of Defense, Washington, D.C., 2009.

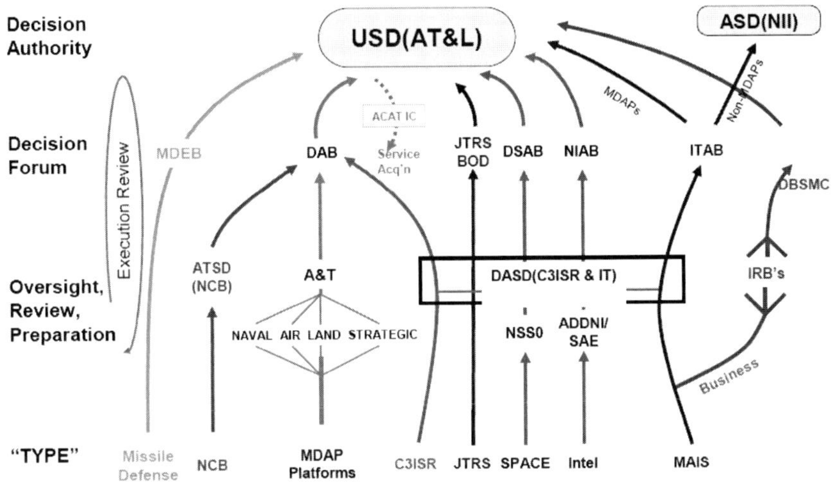

FIGURE A.5 Governance and oversight process flows for various types of programs. SOURCE: Information Technology Acquisition briefing provided by DASD (C3ISR & IT Acquisition), February 25, 2009.

TABLE A.2 Defense Acquisition Board Membership and Advisors for MDAP Program Oversight

DAB Members	DAB Advisors
Under Secretary of Defense (Comptroller)	Principal Deputy USD(AT&L)
Under Secretary of Defense (Policy)	Director, Defense Research & Engineering
Under Secretary of Defense (P&R)	OIPT Leader(s)
Under Secretary of Defense (Intelligence)	Chairman, Cost Analysis Improvement Group
Assistant Secretary of Defense for Networks and Information Integration/DOD CIO	Director, Defense Procurement and Acquisition Policy
Director, Operational Test & Evaluation	Program Executive Officer
Chairman, Program Analysis and Evaluation	Program Manager
Secretaries of the Army, the Navy, and the Air Force	Deputy Under Secretary of Defense (Logistics & Material Readiness)
Director, Acquisition Resources & Analysis	DOD General Counsel
Director, Force Structure (J8)	Deputy Under Secretary of Defense (Industrial Policy)
	DOD Component Acquisition Executives
	Commander, United States Joint Forces Command
	Chair, Functional Capabilities Board(s)

TABLE A.3 Information Technology Acquisition Board Membership and Advisors for MAIS Program Oversight

ITAB Members	ITAB Advisors
Under Secretary of Defense (Comptroller)	Under Secretary of Defense (Policy)
Under Secretary of Defense (P&R)	Under Secretary of Defense (Intelligence)
Deputy DOD Chief Information Officer	Domain Owner
Director, Operational Test & Evaluation	Component CIOs
Deputy DOD Chief Information Officer	Director, Defense Intelligence Agency
Director, Operational Test & Evaluation	Director, Cost Analysis Improvement Group
Chairman, Program Analysis and Evaluation	Representatives of the Joint Staff
Component Acquisition Executives of the Army, Navy, and Air Force	Director, Defense Procurement and Acquisition Policy
DOD Component User Representatives	Director, International Cooperation
Director, Defense Procurement and Acquisition Policy	Deputy Under Secretary of Defense (Logistics and Materiel Readiness)
IT OIPT Lead	Deputy Under Secretary of Defense (Industrial Policy)
Program Executive Officer(s)	Director, Acquisition Resources and Analysis
Program Manager(s)	Deputy Under Secretary of Defense (Installations and Environment)
Cognizant OSD Principal Staff Assistant(s)	
Director, Force Structure (J8)	
DOD General Counsel	
Deputy Director, Developmental Test & Evaluation	

positions and levels of the participants differ, the ITAB has a very similar overall composition, as shown in Table A.3.

Operational test and evaluation (OT&E) to determine the effectiveness and suitability of systems is also a key consideration of the DAMS. Test and evaluation artifacts ranging from strategy documents to final test reports are required for every milestone review. OT&E results are a key factor in the limited-rate initial production and full-rate production decisions. (For software-intensive programs with no production components, Milestone C is a deployment decision.) Prior to achieving a Milestone C production or deployment decision, the system must undergo OT&E under realistic conditions to determine if the threshold requirements have been met and critical operational issues have been satisfied. These threshold requirements come from the approved capabilities development document developed in the JCIDS process. For AIS programs, additional specialized testing is integrated into the OT&E process. This includes interoperability testing and network-ready certification conducted by the Joint Interoperability Test Command (JITC), and information assurance certification and accreditation testing conducted by the designated accrediting authority.[6] For evolutionary programs following the model of Figure A.2, this testing and certification must occur for every evolutionary spiral undertaken by the program.

In short, the DAMS is a complex system of governance and oversight. For any MDAP or MAIS program, preparing for and successfully conducting the series of milestone reviews necessary to deploy capability to end users is a major undertaking.

[6] DODI 8510.01, 2007.

Appendix B

Program Phases and Decision Milestones for SDCI Programs

This appendix and Appendix C provide a somewhat-detailed candidate description of program phases and decision milestones for SDCI and CHSS programs, respectively. Rather than being explicitly prescriptive, these appendixes are meant to offer plausible potential ways in which the committee's recommended changes might be incorporated that align with current acquisition methods. There are, of course, other possible implementations of the committee's recommendations.

MATERIAL DEVELOPMENT DECISION (SDCI PROGRAMS)

Purpose: The purpose of the Material Development Decision (MDD) is to validate the need for material development to address the requirement for a new or improved mission capability as a result of a projected deficiency or obsolescence in existing systems that cannot be addressed appropriately through continued evolution of those systems; a technological opportunity; or an opportunity to reduce operating cost. An additional purpose is to gain approval of a draft top-level ("big-R") capability description and draft concept of operations (CONOPS) for the capability. These documents should each be top-level documents of relatively brief length but should provide the overarching direction for the program.

PM Responsibilities: Not applicable; no program exists at this point.

PMT Responsibilities: Develop and agree on the rationale for the material development need, the draft mission capability description, and the draft CONOPS for the capability.

MDA Responsibilities: Validate the rationale for the material development need, and approve the draft capability description and the draft CONOPS for the capability. If appropriate, resolve any issues not mutually agreed on across the members of the portfolio management team (PMT). Provide guidance to apply during the Business Case Development Phase, as appropriate.

BUSINESS CASE DEVELOPMENT PHASE (SDCI PROGRAMS)

The Business Case Development Phase enables leadership to make an informed, rational initial decision to invest in a program. It should further evolve the draft capability description and draft CONOPS and develop alternative approaches or system concepts for the proposed program. It should formalize the approach to quantify costs that will be incurred in the program and benefits expected to be achieved by the program, and conduct an analysis of the trade-offs among the alternatives to assess the anticipated costs and benefits of each in order to recommend a preferred approach or system concept. It should also identify major risk factors that could jeopardize success and propose mitigation strategies for each major risk factor. In so doing, it should develop a proposed schedule and budget for the capability increments from the initial capability increment through to the final capability increment and anticipated life-cycle costs, and propose an allocation of the top-level requirements identified in the draft capability description to the capability increments. This proposed plan should be a "living document" intended to be refined in subsequent planning and analysis phases as learning and communications continue throughout the multiple capability increments of the program.

The Business Case Development Phase is carried out under the leadership and direction of the PMT for the proposed program.

MILESTONE A: PLANNING, ANALYSIS, AND CONCEPT DEMONSTRATION DECISION (SDCI PROGRAMS)

Purpose: The purpose of the Milestone A: Planning, Analysis, and Concept Demonstration Decision is to validate the business case and analysis of alternatives, and to authorize entry into the initial Planning, Analysis, and Concept Demonstration Phase.

PM Responsibilities: Not applicable; no program exists at this point.

PMT Responsibilities: Develop and agree to the business case and recommended alternative.

MDA Responsibilities: Approve the business case and recommended alternative. Authorize commencement of the initial Planning, Analysis, and Concept Demonstration Phase and provide guidance for that phase.

INCREMENT 1 PLANNING, ANALYSIS , AND CONCEPT DEMONSTRATION PHASE (SDCI PROGRAMS)

The purpose of the Planning, Analysis, and Concept Demonstration Phase is to provide further validation of the recommended alternative approach and its projected costs and benefits prior to formal initiation of the program. Depending on the objectives of the program and the degree to which it is attempting to address requirements or employ technology in a manner without substantial precedent, this phase can use prototyping beneficially to demonstrate key features of the proposed solution, such as the following.

- Demonstrate key concepts that the system is intended to address.
- Demonstrate the approach to key nonfunctional requirements and explain its anticipated feasibility.
- Demonstrate the approach to high-value business or warfighting functional "big-R" requirements and explain its anticipated feasibility.
- Demonstrate the approach to quantifying benefit in terms meaningful to the end users.
- Demonstrate the ability to function properly in the intended operational environment for the system, including constrained communications and networking environments.

Based on the learning and communication with the end users that result from this concept demonstration prototyping, the planning and analysis efforts should refine the draft capability description and CONOPS, refine the business case, and refine the allocation of the top-level requirements in the draft capability description to the capability increments. It should also establish a vision for the end-state target architecture appropriate to support the full scope of intended capability and appropriate to support the full scale of the intended deployment. Care must be exercised, however, to avoid having development of an ideal architecture inappropriately severely delay delivery of meaningful capability to end users. Finally, it should develop the initial proposed "small-r" requirements for the first capability increment and develop an initial allocation of those requirements to the time-boxed iterations within the overall capability increment. In developing this allocation, it should adhere to the principle of focusing on difficult technology issues and high-value business or warfighting functional capabilities in early increments.

If the capability to be developed in each increment of the IID-based program requires complex deployment processes and/or specific end-user training, the planning activities should also address the intended scope and timeline of deployment across the multiple capability increments of the program.

The Increment 1 Planning, Analysis, and Concept Demonstration Phase can readily be adapted to a multiple-award contract model with a subsequent competitive selection of a single provider coincident with the initial Milestone B decision if such additional competition is deemed appropriate in the program's acquisition strategy.

INCREMENT 2 AND BEYOND PLANNING AND ANALYSIS PHASES (SDCI PROGRAMS)

For subsequent planning and analysis phases after the initial one leading to the initial Milestone B Program Initiation Decision, the above process can be substantially abbreviated. Follow-on prototyping should be performed if and only if it adds value to resolve fundamental issues or to enable a choice among alternative approaches (operational or technological) in a manner not readily done within the iterations of the capability increment. Based on learning in the previous capability increments and from field experience with the deployed product, all planning and analysis phases after the initial one should focus on incremental refinement of the following:

- The top-level capability description and CONOPS,
- The business case,
- The allocation of top-level requirements from the capability description for the increment to each of the time-boxed iterations, and
- Development of the initial proposed small-r requirements for the next capability increment and their initial allocation to the time-boxed iterations within the capability increment.

If the capability to be developed in each increment of the IID-based program requires complex deployment processes and/or specific end-user training, the planning activities should also address the intended scope and timeline of deployment across the multiple capability increments of the program.

MILESTONE B: PROGRAM OR CAPABILITY INCREMENT INITIATION DECISION (SDCI PROGRAMS)

Purpose: The purpose of the Milestone B: Program or Capability Increment Initiation Decision is to validate the overall refined capability description and how big-R requirements are allocated across all subsequent increments, and the time-phased scope of deploying capability across the increments. It must also validate the proposed small-r refined requirements allocated to the next increment, together with the plan for how the increment will be executed. If this is not the Milestone B decision for the first increment, the Milestone B decision should ideally coincide with the Milestone C decision for the previous increment.

PM Responsibilities: Conduct the planning and analysis phase and any required concept demonstrations. Together with the PMT, refine the allocation of big-R requirements to all subsequent increments and develop the small-r refined requirements to be undertaken in the next increment. Develop an initial allocation of the small-r requirements for the next increment across the series of iterations that will be conducted in the next increment.

PMT Responsibilities: Together with the PM, develop the refined allocation of big-R requirements to all subsequent increments. Together with the PM, develop and reach agreement on the small-r refined requirements for the next increment.

MDA Responsibilities: Approve the overall program baseline allocation of big-R requirements across all subsequent increments and the small-r requirements allocated to the next increment. Authorize commencement of the program or next-increment System Development and Demonstration Phase and provide any guidance deemed appropriate for the conduct of that phase.

SYSTEM DEVELOPMENT AND DEMONSTRATION (SDCI PROGRAMS)

The purpose of the System Development and Demonstration (SDD) Phase is to develop the next increment of capability through a learning and communicating cycle of time-boxed iterations informed by the end user's perspective and integrated test and evaluation as key components of the learning and communications process throughout the iterations. During each iteration of this learning and communications cycle, the PM, in cooperation with the "voice of the end user," has the ability to refine and reprioritize the small-r requirements for subsequent iterations consistent with the learning that has occurred in prior iterations. If, during this process, the PM determines it will be impossible to substantively meet within the established cost and schedule baseline the big-R requirements allocated to the increment, the PM should notify the PMT and the MDA and present a recommended course of action. Conversely, if the PM, informed by the perspective of end users, determines that intermediate iterations have produced sufficiently significant capability to warrant early fielding, the PM should likewise notify the PMT and MDA with a recommended course of action.

MILESTONE C: CAPABILITY INCREMENT DEPLOYMENT DECISION (SDCI PROGRAMS)

Purpose: The purpose of the Milestone C: Capability Increment Deployment Decision is to assess the risk versus benefit of deploying the capability developed during the SDD phase to the subset of end users within the intended deployment scope. This is a marked departure from the current approach of assessing whether a fixed set of requirements including key performance parameters (KPPs) have been achieved with cost and schedule floating to whatever level is necessary to achieve that objective. In this approach, the increment is time-boxed and executed with the cost and schedule constrained to the baseline set at the previous Milestone B decision and the degree of success in meeting the big-R requirements set for the increment. The attendant risk versus benefit of fielding the product is the key consideration for the decision. If there are subsequent increments, this Milestone C decision should ideally be conducted coincident with the Milestone B decision for the subsequent increment, since many of the factors affecting the deployment decision can also materially affect the composition of the next increment.

PM Responsibilities: In conjunction with operational end users, DT&E and OT&E stakeholders, security C&A stakeholders, and interoperability stakeholders that have been responsible for the integrated test and evaluation conducted throughout the increment, the PM shall assess the risk versus benefit in deploying the capability to the intended end users and make a deployment recommendation to the PMT and the MDA.

PMT Responsibilities: Validate the assessment of benefits and risks conducted by the PM and all integrated T&E stakeholders and their deployment recommendation based on that assessment. Make an independent deployment recommendation to the MDA.

MDA Responsibilities: Approve (or disapprove) the deployment recommendation made by the PM and provide guidance for subsequent capability increments.

DEPLOYMENT PHASE (SDCI PROGRAMS)

The purpose of the Deployment Phase is to deploy the developed capability to the intended subset of end users. If the capability developed during the SDD Phase and its deployment approach are straightforward, the Deployment Phase can be a very simple and straightforward activity (for example, making the capability available online as a service, or over the network as an automatically installable download). If, however, the capability is complex, and especially if there are interdependencies with other programs, complex installation procedures not suitable for "point-and-click" installation by the end user, and/or unique training requirements, significant planning and effort may be required to deploy the capability. This is especially the case for deployable or deploying units because the capability deployment activities of the acquisition program must then be integrated into the overall operational employment schedule for each unit. In that case, capability deployments will generally be planned during the periods between a unit's operational deployments so that training can be conducted prior to the workup process preceding the next operational unit deployment to a forward area of responsibility.

During the Deployment Phase for these more complex cases, the PM must perform the detailed coordination with the operating forces to plan, schedule, and execute capability deployment activities including preparatory work, pre-installation test and checkout specific to each installation, installation, post-installation operational validation, training, and, if appropriate, integration validation with other interdependent capabilities (for example, across a portfolio). In Figure 3.2, the end of the Deployment Phase for an increment is aligned with the Milestone C decision for the next increment since generally, the subsequent increment will represent improved capability and, once it is available, should be utilized for subsequent deployments whenever possible.

OPERATIONS AND SUSTAINMENT PHASE (SDCI PROGRAMS)

The purpose of the Operations and Sustainment Phase is to support all previously deployed versions of a capability still in operational use. This support includes activities such as operating an end-user help desk and responding to problems encountered in operational use of the capability, including the development and distribution of patches and maintaining a configuration status accounting baseline for all installations of the capability. This phase also includes the collection of metrics built into the deployed capability such as data on utilization of specific features, so that unused elements of the capability can be removed.

This phase is shown with an indeterminate duration in Figure 3.2, since the complexities discussed above in conjunction with the Deployment Phase will, in general, result in multiple versions in use at any given point in time. Even in the simplest case, when end-user "point-and-click" automatic download and installation is the deployment approach, a predetermined interval is typically necessary to allow time for the installed base of end users to update their installations prior to terminating support for previously released versions, as is often the case in commercial software applications.

Appendix C

Program Phases and Decision Milestones for CHSS Programs

This appendix and Appendix B provide a somewhat-detailed candidate description of program phases and decision milestones for CHSS and SDCI programs, respectively. Rather than being explicitly prescriptive, these appendixes are meant to offer plausible potential ways in which the committee's recommended changes might be incorporated that align with current acquisition methods. In several cases the program phases and decision milestones for CHSS programs are similar to those for SDCI programs, which are elaborated in Appendix B. There are, of course, other possible implementations of the committee's recommendations.

MATERIAL DEVELOPMENT DECISION (CHSS PROGRAMS)

The purpose of the Material Development Decision (MDD) and the responsibilities of the PM, PMT, and MDA for CHSS programs are the same as for SDCI programs.

BUSINESS CASE DEVELOPMENT PHASE (CHSS PROGRAMS)

The purpose of the Business Case Development Phase for CHSS programs is the same as for SDCI programs, though with much greater emphasis placed on aligning the business strategy and investment strategy with the technical incremental capability strategy as discussed in Appendix B. Correspondingly there should be much less emphasis on a concept of operations (CONOPS) for purely unmodified COTS hardware, software, and services. As with SDCI programs, the Business Case Development Phase is carried out under the leadership and direction of the PMT for the proposed program.

MILESTONE A: PLANNING, ANALYSIS, AND CONCEPT DEMONSTRATION DECISION (CHSS PROGRAMS)

The purpose of the Milestone A: Planning, Analysis, and Concept Demonstration Decision Phase and the responsibilities of the PM, PMT and MDA for this category of IT acquisition programs are conceptually similar to those for SDCI programs. The difference at this decision milestone and in the subsequent program phase is that concept demonstration should be undertaken if and only if there are clear issues or questions that must be resolved through demonstration that cannot be resolved in successive capability increments. This will frequently not be the case for the use of unmodified COTS products or services. As such, concept demonstration should be regarded as optional, with a bias to not performing it for most programs. The principal focus should be on the planning and analysis activities. Due to the nature of CHSS programs, this will substantially be market research-based analysis with planning extended into the Deployment Phase.

INCREMENT 1 PLANNING, ANALYSIS, AND CONCEPT DEMONSTRATION PHASE (CHSS PROGRAMS)

The purpose of the Planning, Analysis, and Concept Demonstration Phase for CHSS programs is similar that for SDCI programs with the exception of the change in emphasis discussed above. Further, requirements will typically focus on capability, capacity, and key nonfunctional requirements (e.g., operational availability and environmental qualification for hardware).

As with the software development and COTS software integration category of IT programs, the Planning, Analysis, and Concept Demonstration Phase for the first increment can readily be adapted to a multiple-award contract model with a subsequent competitive selection of a single provider coincident with the initial Milestone B decision if such additional competition is deemed appropriate in the program's acquisition strategy.

INCREMENT 2 AND BEYOND PLANNING AND ANALYSIS PHASES (CHSS PROGRAMS)

For subsequent planning and analysis phases after the initial one leading to the initial Milestone B Program Initiation Decision, the above process can be substantially abbreviated. The purpose of the Increment 2 and Beyond Planning, and Analysis Phases for CHSS programs is the same as that for the second increment in SDCI programs.

MILESTONE B: PROGRAM OR CAPABILITY INCREMENT INITIATION DECISION (CHSS PROGRAMS)

The purpose of the Milestone B: Program or Capability Increment Initiation Decision and the responsibilities of the PM, PMT, and MDA for CHSS programs are the same as those for SDCI programs.

SYSTEM DEVELOPMENT AND DEMONSTRATION (CHSS PROGRAMS)

As with SDCI programs, the purpose of the System Development and Demonstration (SDD) Phase for CHSS programs is to provide the next increment of capability. Since developmental efforts are not involved, however, the nature of the learning and communications cycle and the role of the end user and other stakeholders change, as does integrated test and evaluation. Since the focus is on COTS software configuration, hardware integration, or hardware ruggedization to meet environmental qualification requirements, and not on software development, time-boxed iterations can still play a role but are not as critical as they are for SDCI programs.

MILESTONE C: CAPABILITY INCREMENT DEPLOYMENT DECISION (CHSS PROGRAMS)

The purpose of the Milestone C: Capability Increment Deployment Decision and the responsibilities of the PM, PMT, and MDA are the same for CHSS programs as they are for SDCI programs, with one addition: validating the attainment of an environmentally qualified first article for COTS hardware programs targeted at deployable units. As SDCI programs, if there are subsequent increments, this Milestone C decision should ideally be conducted coincident with the Milestone B decision for the subsequent increment, since many of the factors affecting the deployment decision can also materially affect the composition of the next increment.

DEPLOYMENT PHASE (CHSS PROGRAMS)

The purpose of the Deployment Phase is the same for CHSS programs as it is for SDCI programs.

OPERATIONS AND SUSTAINMENT PHASE (CHSS PROGRAMS)

The purpose of the Operations and Sustainment Phase for this category of IT acquisition programs is the same for CHSS programs as it is for SDCI programs.

Appendix D

Programs That Succeeded with Nontraditional Oversight

The programs discussed below are examples of large and complex DOD IT programs that were demonstrably successful from an end-user perspective while being executed with tailored, focused, proactive, accountable oversight of the kind advocated in this report.

The **Force XXI Battle Command Brigade and Below (FBCB2)** command-and-control system was developed as a central element of the Army's Advanced Warfighting Experiments (AWE) in the mid-1990s. A companion development was the Army's Tactical Internet, a data communications capability designed by adding commercial router technology to legacy tactical communications devices. The Tactical Internet and FBCB2 formed a capability to disseminate real-time battle command information across the force. A key feature is the ability of the system to generate location information on each FBCB2-equipped vehicle based on Global Positioning System feeds and to automatically distribute this information to all other members of the force equipped with FBCB2. The command-and-control information is automatically updated on digital map displays on weapons platforms and in tactical operations centers. Early in its development stage, FBCB2 was deployed to the Central Technical Support Facility (CTSF) at Fort Hood, Texas, where soldiers from the 4th Infantry Division provided continual feedback on the system design and soldier-machine interfaces.

New capability packages were regularly deployed and evaluated at the CTSF by both the test community and end users, effectively executing an agile development approach. The soldiers functioned as a user jury and provided candid assessments and recommendations. A board of senior Army general officers conducted regular assessments and provided guidance as FBCB2 and related experimental systems were prepared for the capstone digitization experiment at the National Training Center in March 1997. More than 1000 FBCB2 systems were procured and deployed to the 4th Infantry Division for that experiment. Some systems were MILSPEC, some were ruggedized, and some were COTS-based. This approach provided a set of optional configurations that were evaluated by end users and the operational test organization to provide feedback to the Army on the performance of the configurations. Note that this feedback was not a test-fail evaluation with a report 120 days after the field event that is typical in the formal test environment; rather, the testers provided both daily feedback and an early capability assessment wrap-up that the Army used to make a "best value" determination (the answer ultimately was the ruggedized COTS variant). Following the 1997 AWE, the Army used the feedback from end users to make changes and enhancements to the FBCB2 system. The test articles remained with the 4th Infantry Division for training and further development of tactics, techniques, and procedures for operational use. In 1998 an operational evaluation (Limited User Test) was conducted, and low-rate initial production of 6000 FBCB2 systems was authorized to field the capability to the 4th Infantry Division and 1st Cavalry Division. Today approximately 40,000 FBCB2 systems are in operational use in the Army and the Marine Corps. Moreover, the FBCB2 system is the baseline for a follow-on variant named the Joint Battle Command Platform. The FBCB2 system was recognized as one of the five best-managed software programs in the entire U.S. government and was awarded the *Federal Computer Week* Monticello Award (given in recognition of an information system that has a direct, meaningful impact on human lives). FBCB2 exemplified the type of decentralized agile development approach that this report recommends.

The **Blue Force Tracker (BFT)** is a variant of FBCB2 that uses satellite-based communications in lieu of the terrestrial communications capabilities used in FBCB2. Early variants were deployed on surrogate commercial computers for use during the conflict in the Balkans in the late 1990s. During 2002 an intensive effort was initiated with supplemental funds to develop and deploy BFT for forces being prepared for Operation Iraqi Freedom. Contractors and Army program managers were deployed to Kuwait, where BFT was installed on weapons platforms and soldiers were trained in its use. Since the baseline FBCB2 program was in the production and deployment stage, the infrastructure for that program was leveraged to execute BFT very rapidly without burdensome oversight. This is a prime example of an opportunity to bring capabilities to warfighters rapidly by leveraging a modification-in-service approach to streamline the front end of the acquisition process and execute the program in a highly decentralized manner. Programs that are in production are ideally positioned to adopt the agile processes that this report recommends as a channel to acquire and field new capabilities for warfighters by adaptation or technology insertion, without incurring the time-consuming processes of new starts.

The **Joint Network Node (JNN)** also used the opportunity to leverage a modification-in-service funding line. In this case the funding line had been in place for many years as part of the Army's Mobile Subscriber Equipment (MSE) program. Its annual funds paid for multiple incremental developments and fieldings of capabilities over MSE's decades-long life cycle. MSE started with ACAT I-level oversight at the OSD level, but oversight authority subsequently was delegated to the Army. Despite its regular upgrades through technology insertion, MSE in its current configuration could no longer provide adequate support to the deployed warfighters in Operation Iraqi Freedom (OIF) because it lacked the mobility and broadband satellite-based capabilities needed by warfighters deployed in widely dispersed locations. The Warfighter Information Network-Tactical (WIN-T), an ACAT ID program in the development phase that will eventually resolve current communications deficiencies, was years away from fielding when the deployed warfighters identified an urgent need for significantly enhanced capabilities. Consequently, the Army allocated supplemental funding and used the existing MSE modification-in-service contract to initiate the JNN program and respond rapidly to a CENTCOM urgent operational needs statement requesting communications capabilities better than what their deployed MSE could deliver. A solution to meet the requirement was designed using COTS and government off-the-shelf capabilities, and a relationship was established with the 3th Infantry Division so that the new JNN capability could be deployed and training completed before its OIF rotation. The program proved to be so successful that a decision was made to deploy additional JNNs to other divisions preparing to deploy. These fieldings were unencumbered by the formal process of a program of record (POR) because the success metrics were "good, fast, and affordable." Eventually, the JNNs became so widespread and the funding level accumulated to so high a level that some of the standard acquisition processes were appropriate. As a result, an initial operational test and evaluation was conducted in-theater and the program was ultimately melded into the WIN-T program. In sum, this program proved to be agile and responsive to warfighter needs. Lessons learned from experience with this program can be applied as the DOD adopts a more responsive process for the acquisition of IT-based systems.

The **Command Post of the Future (CPOF**) is a command-and-control program built with advanced visualization and collaboration technology from the commercial and academic sectors that was initiated by DARPA. There was early collaboration with the user community during system development, and the system was deployed for evaluation, training, and interoperability enhancements at the CTSF at Fort Hood, Texas. CPOF interoperates with the Army's command-and-control POR. It received high praise from the end users, who requested that it be deployed to OIF. Based on its success the system was transitioned into a formal POR for support, further fielding, and upgrades to meet evolving end-user requirements. This is another example of a program that thrived in the absence of formal ACAT I-level program oversight. Partnering of this kind between DARPA and the Services and agencies can be a means to achieve agile development and shorten the front end of the IT system acquisition process.

The **Tactical Ground Reporting System (TIGR)** is another DARPA program. It is a multimedia reporting system for soldiers at the patrol level, allowing users to collect and share information to improve situational awareness and to facilitate collaboration and information analysis among junior officers. It is based on commercial information technology and was developed using rapid and agile acquisition processes without going through the normal oversight process. It was developed in collaboration with end users and has evolved into a highly valued, widely deployed system in Iraq and Afghanistan. Like CPOF, this program should be evaluated in depth for lessons learned that can be deployed across the DOD IT system acquisition community.

Appendix E

Briefings to the Committee

JUNE 30, 2008

Lt. Gen. Charles E. Croom, Jr., DISA—*Opening Remarks*
Steven Hutchison, DISA—*Charge from the Sponsor*
Steven Hutchison, DISA—*Testing and Evaluation for Information Technology*
Martin Gross, DISA—*View from the Component Acquisition Executive*
Becky Harris, DISA—*Net-Centric Enterprise Services*
Dave Bennett, DISA—*Program Executive Office for Command and Control Capabilities*
Robert Gorman, Mark Orndorf, Luanne Overstreet, Jimaye Sones—*DISA Panel*
Dan Sturman, Google—*Beta Testing of Google Services*
Dave Aland, Wyle—*Evaluating 1A, Measuring More Than Failure*
Timothy J. Harp, Deputy Assistant Secretary of Defense (C3ISR&IT Acq)

JULY 1, 2008

Lt. Gen. (ret.) Ronald Kadish, Booz-Allen—*Defense Acquisition Performance Assessment (DAPA) Study*
William Johnson, Program Executive Office of Integrated Warfare Systems—*ARCI-A Historical Perspective*

AUGUST 12, 2008

Martin Gross, DISA Component Acquisition Executive (by telephone)

SEPTEMBER 11, 2008

Tony Montemarano, DISA Component Acquisition Executive—*GIG Bandwidth Expansion and DOD Acquisition*
John Garing, DISA Chief Information Officer
Randy Hite, Government Accountability Office—*Overview of GAO's Report on Global Combat Support System—Marine Corp*
Nancy Spruill, OSD—*Defense Acquisition from a Management Perspective*

SEPTEMBER 12, 2008

Martin Westphal and Alex Urrutia, Joint Force Command—*Command and Control Capability Portfolio Management*
Mike Krieger, Deputy CIO, United States Army
Robert Gorman, DISA General Counsel
Mark Drapeau, National Defense University

DECEMBER 8, 2008

Jacques Gansler, Former Under Secretary of Defense for Acquisition, Technology and Logistics—*DOD I.T. Acquisition*
Steve Kelman, Former Director, Office of Federal Procurement Policy
John Goodenough, committee member—*Site Visit Report*
Bruce Amato, Office of the Under Secretary of Defense Acquisition and Technology—*DOD Software and Systemic Issues and Recommendations*
David Wennergren, DOD Deputy Chief Information Officer
Stuart Starr, National Defense University—*Actions to Enhance the Use of Commercial Information Technology in DOD Systems*
Don Johnson, Defense Science Board—*Challenges in Acquisition Technology*
John Stenbit, Former Assistant Secretary of Defense for Networks and Information Integration

DECEMBER 9, 2008

Ron Jost, OSD—*Network Centric Capability Portfolio*

JANUARY 30, 2009

John Landon, Vice President-Missiles, Technology and Space Programs, Northrop Grumman (former Deputy Assistant Secretary for Command, Control, Communications, Intelligence, Surveillance, Reconnaissance and Information Technology Acquisition in OSD)

Priscilla Guthrie, Director of the Information Technology and Systems Division, Institute for Defense Analyses (former Deputy Assistant Secretary, Office of CIO)

FEBRUARY 24, 2009

Keith Seaman, Defense Business Systems Acquisition Executive—*Tranformational Times: Facing the Challenges* (teleconference)

Mike Dettman, U.S. Navy—*PEO C4I Program*

Dan Sturman, Google—*Agile Development with Large Teams*

FEBRUARY 25, 2009

Timothy J. Harp, Deputy Assistant Secretary of Defense (C3ISR&IT Acq)—*A New Model for IT Acquisition in DOD*

Appendix F

Biosketches of Committee and Staff

COMMITTEE MEMBERS

William H. Campbell, *Co-Chair*, is vice president, Advanced Network Systems, BAE Systems, Inc. He joined BAE Systems in 2002 and established the Information and Communication Networks Business Area, which he led as vice president and general manager until 2007. In that capacity he provided systems-level solutions for warfighters. Prior to joining BAE Systems, he was the University of California's chief information officer (CIO) and associate vice president, information resources and communications. He served in the office of the president of the university system with responsibility extending through ten campuses, five medical centers, and three national laboratories. His duties included implementing the university's New Business Architecture, overseeing the Digital California Project, and serving on the board guiding the deployment of Internet-2 in California. Mr. Campbell retired from the Army at the rank of Lieutenant General. His 38-year career as a soldier culminated with duty as the Army's director of information systems for command, control, communications and computers (G6); as CIO for the U.S. Army; and as a military deputy to the Army acquisition executive. During his military career, he held operations and military intelligence positions, including command from company through brigade. As a general officer, he held positions in information management, research and development, and systems acquisition, including 10 years in program executive officer jobs. He represented the U.S. Army on NATO R&D committees, led the campaign to improve computer security, initiated a biometric identification program, directed the Advanced Precision Strike Demonstration Program, and was the systems architect for the advanced warfighting experiments that transformed the Army to a digitized force. He currently serves on the Army Science Board and two Defense Science Board panels. He is a past member of the federal and DOD CIO Councils, DOD's Military Communications-Electronics Board, Microsoft's Global Executive Roundtable, Dell's Platinum Council, the Bay Area Regional Technology Alliance, the National Science Center Advisory Board, the Corporation for Education Network Initiatives in California, and the California Information Technology Commission. In addition he served previously as a member of the NRC Committee on Strategies for Network Science, Technology, and Experimentation. Mr. Campbell is a graduate of the Army's Command and General Staff College and the Naval War College. He earned an MBA with a computer science concentration from Texas Tech University.

Dawn C. Meyerriecks,[1] *Co-Chair,* has provided senior leadership business and technology consulting direction to government and commercial clients. This includes competitive intelligence and landscape, product and service futures and marketability intersection, smart sourcing, and evolving technical and business best practices. In addition to consulting, she serves on a number of government and commercial advisory boards, including the STRATCOM C2 Advisory Group, the NSA Advisory Board, the Defense Science Board, Cranite Advisory Board, and the SunFed Advisory Board. From 2000 to 2006, Ms. Meyerriecks served as the senior vice president for product technology at AOL. While at AOL, she was responsible for full life-cycle development and integration of all consumer-facing AOL products and services, including the relaunch of aol.com, AOL Instant Messenger, and the open client platform. Prior to AOL, she worked for nearly 10 years at the Defense Information Systems Agency (DISA), where she was the chief technology officer and technical director for the Joint Interoperability and Engineering Organization (JIEO). Her last assignment was to charter and lead a new Global Information Grid (GIG) Enterprise Services organization. Ms. Meyerriecks worked at the Jet Propulsion Laboratory as a senior engineer and product manager before her tenure at DISA. In addition to being named the Government Computer News Department of Defense Person of the Year for 2004, Ms. Meyerriecks has been honored with numerous other awards, including InfoWorld 2002 CTO of the year; Federal Computer Week 2000 Top 100; InfoWorld 2001 CTO of the year for the government sector; the Presidential Distinguished Service Award, November 2001; the Senior

[1] Dawn Meyerriecks resigned from the committee in September 2009 upon her appointment as Deputy Director of National Intelligence for Acquisition and Technology.

Executive Service Exceptional Achievement Awards in 1998, 1999, 2000; and the National Performance Review in August 1996. In November 2001, she was featured in *Fortune* magazine as one of the top 100 intellectual leaders in the world. She earned an M.S. in computer science from Loyola Marymount University.

Robert F. Behler is the deputy general manager and senior vice president in the Command and Control Center at the MITRE Corporation. The center serves MITRE's Department of Defense sponsors and focuses on creating a joint command, control, and communications system. Mr. Behler leads the center's work for Department of Defense sponsors. Before joining MITRE in April 2006, Mr. Behler was general manager of Precision Engagement at Johns Hopkins University's Applied Physics Laboratory. In this position he supervised more than 250 scientists and engineers working on advanced command, control, intelligence, surveillance, and reconnaissance (C2ISR) programs for the Department of Defense. Under Mr. Behler's leadership, the Precision Engagement organization turned new and emerging technologies into transformational operational capabilities. Mr. Behler retired from the Air Force as a major general in 2003. During his distinguished 31-year career, he accumulated extensive experience in test and evaluation and developing advanced command, control, communications, computers, intelligence, surveillance, and reconnaissance (C4ISR) technologies at all levels. He was an experimental test pilot and has flown more than 65 aircraft, including the SR-71 and U-2 aircraft. Before retiring, Mr. Behler was commander of the Air Force C2ISR Center at Langley Air Force Base, where he was principal C2ISR advisor to the Secretary and Chief of Staff of the Air Force. Prior to that, he served as deputy commander of NATO Joint Headquarters North in Stavanger, Norway, and was the senior U.S. military officer in Scandinavia. He has also served as director of command, control, communication, computers, and intelligence at the U.S. Strategic Command at Offutt Air Force Base and as chief of the U.S. Air Force Senate Liaison Office. Mr. Behler entered the Air Force in 1972 as a distinguished graduate of the Air Force Reserve Officer Training Corps program at the University of Oklahoma. He is an associate fellow of the Society of Experimental Test Pilots and a member of the Armed Forces Communications and Electronics Association. Mr. Behler currently serves on the NAS Committee on Advancing Software-Intensive Systems Producibility. He earned an M.S in aerospace engineering and an MBA and was a National Security Fellow at the John F. Kennedy School of Government, Harvard University.

Philip E. Coyle III is a senior advisor to the president of the World Security Institute, and to its Center for Defense Information, a Washington, D.C.-based national security study center. He is a recognized expert on U.S. and worldwide military research, development and testing; on operational military matters; and on national security policy and defense spending. In 2005 and 2006, Philip Coyle served on the nine-member Defense Base Realignment and Closure Commission, appointed by President George W. Bush, and nominated by House Democratic Leader Nancy Pelosi. The commission was responsible to determine those U.S. military bases and facilities to be closed or realigned beginning in late 2005. Beginning in late 2004, Mr. Coyle served on Governor Arnold Schwarzenegger's Base Support and Retention Council, from which he resigned to serve on the President's Commission. From September 29, 1994, through January 20, 2001, Mr. Coyle was assistant secretary of defense and director, Operational Test and Evaluation, in the Department of Defense, and is the longest serving director in the 20-year history of the office. In this capacity, he was the principal advisor to the Secretary of Defense on test and evaluation in the DOD. At the DOD, Mr. Coyle's responsibilities included stewardship of the major range and test facility bases of the DOD, including the large test ranges and test centers that the DOD operates from Maryland and Florida to California and Hawaii. As director, Operational Test and Evaluation, Mr. Coyle had responsibility for overseeing the test and evaluation of more than 200 major defense acquisition systems. This included reporting to the Secretary of Defense and to Congress on the adequacy of the DOD testing programs, and on the results from those testing programs. Mr. Coyle was called on regularly to testify before Congress and to brief congressional staff on the status of major defense acquisition programs. Mr. Coyle has more than 40 years of experience in research, development, and testing matters. From 1959 to 1979, and again from 1981 to 1993, Mr. Coyle worked at the Lawrence Livermore National Laboratory in Livermore, California. From 1987 to 1993, he served as laboratory associate director and deputy to the laboratory director. In recognition of his 33 years of service to the Laboratory and to the University of California, the university named Mr. Coyle Laboratory Associate Director Emeritus. During the Carter administration, Mr. Coyle served as principal deputy assistant secretary for defense programs in the Department of Energy. In this capacity he had oversight responsibility for the nuclear weapons testing programs of the department. Currently he is serving on the National Research Council Standing Committee on Biodefense at the U.S. Department of Defense, and recently he served on two National Research Council studies of biological agent detection and identification systems. Mr. Coyle graduated from Dartmouth College with an M.S. in mechanical engineering (1957) and a B.A. (1956).

Renato A. DiPentima served as president and chief executive officer of SRA International from January 2005 through March 2007. Prior to assuming this position, he served as president and chief operating officer. He was initially an SRA vice president and chief information officer (CIO). During DiPentima's tenure at SRA, he helped the company grow from $135 million in revenue to $1.2 billion. Before joining SRA in 1995, DiPentima was deputy commissioner for systems at the Social Security Administration (SSA), overseeing and managing all information processing, data, and voice communications systems. He chaired the Federal Information Technology Acquisition Improvement team as part of the President's National Performance Review initiatives. He also chaired the Industry Advisory Council's CIO task force, making recommendations to the Federal CIO Council on the roles and responsibility of the new federal CIO. DiPentima is a sought-after speaker on topics dealing with CIO functions and activities, procurement reform, systems modernization, automation, and business process reengineering. He has received many awards, including two presidential rank awards (distinguished and meritorious service). He was selected by *Government Computer News* as the Industry Executive of the Year in 2000 and the Government Executive of the Year in 1993, and was honored as Executive of the Year by the Federation of Government Information Processing Councils in 1995. In 2003, DiPentima was selected by *Federal Computer Week* to the Federal 100 for a fifth time, and he also won its prestigious Eagle Award as Industry Executive of the Year. Also in 2003, he was recognized as the Industry Executive of the Year by the Federal CIO Council, which presented him with an Azimuth Award. DiPentima and SRA Founder and Chairman Ernst Volgenau received the Ernst & Young Entrepreneur of the Year® 2006 Master Award in Greater Washington. In 2006, DiPentima received the American Council for Technology/Industry Advisory Council Janice K. Mendenhall Spirit of Leadership Award in recognition of his significant contributions to the federal information technology community, from improving communications to professional mentoring. He earned a Ph.D. from the University of Maryland.

John M. Gilligan is president of the Gilligan Group, Inc. Prior to his current position he was a senior vice president and director, Defense Sector, at SRA International, Inc. Mr. Gilligan has more than 25 years of managerial experience in leading large information technology organizations. He has expertise in business strategy, organization growth, organizational innovation, financial management, program implementation, and IT security. Mr. Gilligan has served as a chief information officer for the United States Air Force and the U.S. Department of Energy. He is a member of the Cyber Security Commission (formed to advise the 44th President) and the Army Science Board. He also serves on the board of directors for the Center for Internet Security, Hunter Defense Technologies, Inc., Systems and Software Productivity Consortium, and the Armed Forces Communications and Electronics Association. Mr. Gilligan has been a recipient of the Distinguished Civilian Service Medal, Joint Chiefs of Staff, Distinguished Executive Presidential Rank Award, and Meritorious Executive Presidential Rank Award, to name a few. He earned an M.B.A. in finance from Virginia Tech University.

John Goodenough works at Carnegie Mellon University's Software Engineering Institute (SEI). He joined the institute in 1986. He is an SEI Fellow and a fellow of the Association for Computing Machinery (ACM). He was chief technical officer (CTO) of the SEI for several years and is now leading a major new SEI research project on software assurance. This project, which started in October 2007, is investigating problems and solutions for assuring critical properties of large, complex systems of systems. Among the activities conducted under this project was a set of interviews with test and evaluation personnel and systems of systems developers to gain insight into the nature of large-system test and evaluation problems. Dr. Goodenough is also leading a research project applying new assurance concepts (assurance cases) to plug-and-play medical devices. Dr. Goodenough previously was the leader of the SEI's Performance Critical Systems Initiative, a project focused on the assurance of real-time embedded systems through quantitative architectural modeling. The resulting approach is beginning to be used by large aerospace companies both in the United States and in Europe. In recent years, Dr. Goodenough has worked with a number of major systems, in particular, the Army's Future Combat Systems (FCS) and NASA's Constellation project. He earned a Ph.D. from Harvard University in 1970 and an M.A. and an A.B. degree in 1962 and 1961, also from Harvard.

Paul J. Kern serves as a senior counselor for the Cohen Group. In November 2004, Gen. Paul Kern concluded his more than 40-year career in the United States Army when he retired as Commanding General, Army Materiel Command (AMC). In that capacity, and earlier as Commander of the 4th Infantry Division (Mechanized), Gen. Kern left his impact on the Army's future as he led a drive to digitize and transform its warfighting capabilities. With a staff of more than 50,000 civilians and active military members, he won wide respect for his efforts to direct supply-chain improvements, maintain field readiness, and modernize weapons systems throughout the Army while still controlling costs. In June 2004, Gen. Kern undertook a vastly different responsibility when then-Secretary Rumsfeld tapped him to lead the military's internal investigation into the abuses at the Abu Ghraib prison in Iraq. Prior to his command at AMC, he served as the military deputy to the assistant secretary of the Army for acquisition, logistics and technology and was the senior military advisor to the army acquisition executive and the Army Chief of Staff on all research, development, and acquisition programs and related issues. He supervised the Program Executive Officer system and served as the director of the Army Acquisition Corps. Gen. Kern's career has also had stops in the Secretary of Defense office in Washington and several field units. As the senior military assistant to then-Secretary of Defense William Perry, Gen. Kern ensured that the secretary's guidance was implemented throughout the department and in the handling of the most sensitive decisions for the secretary. During that tenure he traveled with Secretary Perry to more than 70 countries, meeting numerous heads of state, foreign ministers, and international defense leaders. He is a member of the board of directors of COVANT Technologies, LLC, and iRobot Corporation. Gen. Kern was commissioned as an Armor lieutenant following graduation from the U.S. Military Academy at West Point in 1967. In 2007 he was elected to the National Academy of Engineering for bringing modern digitization technology to bear on military effectiveness, training, and procurement. He holds master's degrees in both civil and mechanical engineering that he earned in 1973 from the University of Michigan.

H. Steven Kimmel serves as corporate vice president for Alion Science and Technology. He is responsible for their strategic plans and implementation to achieve Alion growth. He leads Alion's management of federal, state, local, and commercial opportunity tracking, capture plans, bid review, and proposal preparation development activities. Prior to joining Alion—a 3600 employee-owned, $800 million professional engineering services company—he was vice president of corporate development, Illinois Institute of Technology Research Institute (2000-2002), where he assessed and devised federal marketplace penetration strategies. At TRW (1998-2000), he was the vice president of business development for the Systems and Information Technology Group, Information Technology and Services Division. There he implemented market capture strategies for C4ISR; logistics, supply and maintenance; test and evaluation; and mission, weapons, and force structure operational analysis principally for Defense Department customers. He began his private-sector endeavors in 1993 at BDM Federal as vice president and assistant business unit general manager for test and evaluation. During the 5 years with BDM he was engaged in company-wide system and operational effectiveness analysis programs that included modeling and simulation of military weapons and C4I systems, logistics (wholesale supply, ammunition, and maintenance operations) and automated (business) information systems in support of the Office of the Secretary of Defense, Joint Staff, military departments, defense agencies, and commercial clients. During his federal civil service career he achieved Senior Executive Service Level 5 status. During the period 1985 through 1993, he served on the Defense Acquisition Board (DAB) and the Major Automated Information System Review Council (MAISRC) at the Office of the Secretary of Defense. As such he advised the Secretary of Defense on matters affecting major DOD program development and production matters. His OSD positions included deputy director, defense research and engineering (plans and resources); deputy director, acquisition policy and program integration; deputy director, test and evaluation; and deputy under secretary of defense acquisition (systems evaluation). He earned a doctorate of science from George Washington University in 1983.

Deidre A. Lee, executive vice president of federal affairs and operations, Professional Services Council, served for 32 years in various positions in numerous federal agencies. She retired from the position of director of management and chief acquisition officer for the Federal Emergency Management Agency, Department of Homeland Security in March 2008. Her responsibilities at FEMA included oversight and management of six of FEMA's lines of business: the Offices of Human Resources, Information Technology, Procurement, Facilities, Security, and Disaster Workforce. Before joining FEMA, Ms. Lee served in the General Services Administration's Federal Acquisition Service (FAS) as assistant commissioner of integrated technology services, providing FAS technology and professional services offerings to customer agencies. From 2000 through 2005, she was the director of defense procurement and acquisition policy at the Department of Defense, where she was responsible for department-wide acquisition and procurement policy matters. Ms. Lee also served in the presidentially appointed, Senate-confirmed position of administrator of the Office of Federal Procurement policy in the Office of Management and Budget and as the National Aeronautics and Space Administration's assistant administrator for procurement. Ms. Lee holds a master's degree in public administration from the University of Oklahoma.

Joshua S. Levine is the chief executive officer of ESP Technologies Corporation, a rapidly growing financial technology and solutions provider to the largest global buy-side financial institutions. Previously, as the chief technology, operations and customer service officer of E*Trade Financial Corp., he was responsible for servicing its banking and brokerage customers. He has been a managing director at Morgan Stanley and at Deutsche Bank. Mr. Levine is a member of several corporate boards, including Securify, Xceedium, and Logical Information Machines. He is a former board member of Archivas, purchased by Hitachi and StorageApps, and then purchased by Hewlett-Packard. Mr. Levine is a board member of the nonprofit DonorsChoose.org and an advisory board member to the U.S. National Counterterrorism Center. He is a former board member of the Georgia Technology Authority. Mr. Levine is the recipient of many technology industry awards and an honorary doctorate. He is the co-author of *Application Systems in APL*, published by Prentice-Hall.

Nachiappan Nagappan works on empirical software engineering and measurement (ESM) at Microsoft Research and is based in Microsoft's Redmond, Washington, research facility. Prior to his current position he earned a Ph.D. in computer science from North Carolina State University in 2005. Dr. Nagappan's research interests are in the field of software reliability, software measurement and testing, and empirical software engineering. He has also worked on social factors in software engineering, aspect-oriented software development, and computer science education. Currently his research focuses on the application of software measurement and statistical modeling to large software systems. He works on the MetriZone project that is targeted at making early estimates of software quality to predict postrelease failures, and is currently focused on the next-generation Windows operating system (Vista). Dr. Nagappan is also working with the WinSE team in the Windows Core Operating Systems Division building next-generation change, risk, and impact analysis tools. His tools have been used in product teams such as Windows Mobile for risk analysis and test prioritization. His research work has also commercially shipped as part of the Visual Studio Team System 2005 and 2008 releases.

Frank A. Perry is the chief technology officer and chief systems engineer for Science Applications International Corporation's Defense Solutions Group, and is also a senior vice president. As the senior technical authority across the group, Dr. Perry is responsible for technology leadership and engineering oversight for all programs in the CMMI® Maturity Level 5 Group of more than 12,000 employees, whose charter spans system engineering and integration, command, control and communications; mission systems; modeling, simulation, and training; and enterprise systems and services. Prior to his current position Dr. Perry was the chief technology officer of the Department of Veterans Affairs, where he was responsible for developing the department's first-ever enterprise architecture. He was the driving technical force behind the consolidation of more than 30 independent networks into an integrated enterprise network, the department's Enterprise Cyber Security Infrastructure program, and rationalization of the department's major processing centers to include electronic vaulting and continuity of operations. From 1998 to 2001 Dr. Perry served as the technical director of the Navy's Space and Naval Warfare Systems Command (SPAWAR), and from 1995 to 1998 he was the technical director of the Defense Information Systems Agency (DISA). At SPAWAR he was a key technical leader behind execution of the Navy's IT-21 initiative, which during his tenure installed broadband IP connectivity, LAN infrastructure, and C4I and combat support computing infrastructure across the fleet. At DISA Dr. Perry was a key technical leader in the development of the Global Command and Control System (GCCS) and took the program in 22 months from inception to worldwide deployment, and shutdown of the major legacy World Wide Command and Control System (WWMCCS), which had been entrenched since the mid-1970s. He also was a key architect behind the Defense-In-Depth Information Assurance approach adopted across the DOD, and he personally drove the initiation of the DOD public key infrastructure, the largest in existence. Prior to federal service as a senior executive Dr. Perry was a partner in several engineering services firms and served in the U.S. Navy as an engineering duty officer. Dr. Perry is a member of the Institute of Electrical and Electronics Engineers (IEEE), the Armed Forces Communications and Electronics Association (AFCEA), the National Defense Industrial Association (NDIA), the Association for Enterprise Integration (AFEI), and the International Council on System Engineering (INCOSE). Dr. Perry holds a Ph.D. in electrical engineering with a minor in computer science from the Naval Postgraduate School.

Vaho Rebassoo is the chief technology officer for the Boeing Company's Shared Services Group. He has more than 30 years of experience in systems engineering and technical management in network and computing. This includes key roles at the Pentagon Telecommunications Center, at Bell Telephone Laboratories, and at Boeing, designing, implementing, and operating large complex networks and computing infrastructures. He joined Boeing in 1984 as chief engineer for the Boeing Telephone Service Modernization Program. He assumed responsibility for all network operations in 1988 and for network technical services in 1992. In 2000, he was assigned responsibility for computing technical services enterprise-wide. In his current role at Boeing he is responsible for strategic planning and direction for computing infrastructure technology in the Boeing Company. Dr. Rebassoo is a member of numerous boards of directors and executive advisory boards, including the Washington Technology Alliance Board, the Department of State Telecommunications Advisory Committee, the UCLA Wireless Research Council, and the Pacific Institute for Mathematical Sciences Board. He earned a Ph.D. in mathematics from the University of Washington in 1977.

Daniel C. Sturman is an engineering director at Google, Inc. At Google he is leading the development of software infrastructure that enables Google applications to operate and scale across massive distributed systems. Areas of focus include storage systems, data systems, Web search engines, networking, and cluster management. Prior to joining Google, he held several technical and managerial positions at IBM. Most recently, he was director, development for DB2 on Linux, Unix, and Windows in IBM's Information Managment Division. Products developed by his team include DB2, DB2 Data Warehouse Edition, and DB2 Alphablox. In this role, he was responsible for timely and quality release of these products including DB2 "Viper" v9 and the first DB2 Data Warehouse Edition (v 9.1). Before joining the DB2 team, he was director for emerging technologies in the IBM Software Group, where he ensured that future technical trends were captured within IBM products, directing research and incubation efforts for the Software Group. In particular, Sturman focused on helping IBM's customers successfully implement Service-oriented architectures through an enterprise service bus approach and drove the vision behind the WebSphere ESB. He started at IBM as a researcher at the IBM T.J. Watson Research Center, where his research focused on revolutionizing the way people build and use distributed systems. His research concentrated on technologies for enterprise messaging and utility computing. Sturman's work on enterprise messaging systems addressed the scalability, performance, and availability of content-based publish/subscribe systems. This work helped form the basis for IBM's WebSphere Business Integrator Message Broker. His work on the Gryphon system broke significant new ground in the scale, performance, and functionality of publish/subscribe systems supporting wide-area networks. His research in computing utilities focused on enabling the dynamic provisioning of complete services over the Internet, to reduce the cost of ownership, provide solution availability, and maintain guaranteed service levels. He earned a Ph.D. in computer science from the University of Illinois at Urbana-Champaign.

CSTB STAFF

Jon Eisenberg is director of the Computer Science and Telecommunications Board of the National Academies. At CSTB, he has also been study director for more than a dozen major studies, including a series of reports exploring Internet and broadband policy and networking and communications technologies. From 1995 to 1997 he was a AAAS Science, Engineering, and Diplomacy Fellow at the U.S. Agency for International Development, where he worked on technology transfer and information and telecommunications policy issues. Dr. Eisenberg received his Ph.D. in physics from the University of Washington in 1996 and a B.S. in physics with honors from the University of Massachusetts at Amherst in 1988.

Kevin Lewis is a senior program officer and study director at the Board on Infrastructure and the Constructed Environment. He has served as a study director on a diverse body of work that includes a study addressing the challenge of aging avionics for the Air Force and the issue of emerging technologies within the facilities asset management domain. His career includes experience within business development and technology policy formation in the information technology services industry. He co-authored a book on open systems in his capacity as the co-chair of an ANSI/IEEE standard effort addressing open systems standards development. He received his bachelor's of science degree from the United States Military Academy at West Point, New York, and his master's in business management from Central Michigan University in Mt. Pleasant, Michigan.

Lynette I. Millett is a senior program officer and study director at the Computer Science and Telecommunications Board, National Research Council of the National Academies. She currently directs several CSTB projects, including a comprehensive exploration of sustaining growth in computing performance and an examination of how best to develop complex, software-intensive systems in the DOD environment. She served as study director for the CSTB reports *Social Security Administration Electronic Service Provision: A Strategic Assessment* (August 2007) and *Software for Dependable Systems: Sufficient Evidence?* (May 2007). Millett's portfolio includes significant portions of CSTB's recent work on software, identity systems, and privacy. She directed the project that produced *Who Goes There? Authentication Through the Lens of Privacy*, a discussion of authentication technologies and their privacy implications; and *IDs—Not That Easy: Questions about Nationwide Identity Systems*, a post-9/11 analysis of the challenges presented by large-scale identity systems. She has an M.Sc. in computer science from Cornell University, where her work was supported by graduate fellowships from the National Science Foundation and the Intel Corporation; and a B.A. with honors in mathematics and computer science from Colby College, where she was elected to Phi Beta Kappa.

Renee Hawkins is the financial and administrative manager for the Computer Science and Telecommunications Board. Since 1990, she has been responsible for the financial management of the board. Ms. Hawkins' longtime, hands-on fiscal management experience includes detailed tracking of costs for as many as 15 projects in progress simultaneously, financial reporting, and contract administration. Prior to joining CSTB, Ms. Hawkins provided administrative support to the NRC's Water Science and Technology Board. She has been with the National Academies since 1984. Ms. Hawkins is currently pursuing a B.A. degree in finance and economics at the University of Maryland/Prince Georges' Community College Alliance Program, where she maintains a position on the Dean's List.

Morgan Motto, a program associate with CSTB from December 2007 until April 2009, supported several projects. Previously, she worked with the Board on Environmental Studies and Toxicology (BEST). Prior to coming to the NRC, Ms. Motto worked as a project manager for international affairs and technology at the U.S. Pan Asian American Chamber of Commerce. She earned a B.A. in international affairs and East Asian studies from the Elliott School of International Affairs at George Washington University.

Virginia Bacon Talati is a program associate for the Computer Science and Telecommunications Board of the National Academies. She formerly served as a program associate with the Frontiers of Engineering program at the National Academy of Engineering. Prior to her work at the Academies, she served as a senior project assistant in education technology at the National School Boards Association. She has a B.S. in science, technology, and culture from the Georgia Institute of Technology and an M.P.P. from George Mason University with a focus in science and technology policy.

Appendix G

Acronyms

3CE	Cross Command Collaborative Effort
ACAT	acquisition categories
AF ICE	Air Force Integrated Collaborative Environment
AFOTEC	Air Force Operational Test & Evaluation Center
AIS	automated information system
AoA	analysis of alternatives
ASD	agile software development
ASD NII	Assistant Secretary of Defense for Networks and Information Integration
AT	acceptance team
ATEC	Army Test & Evaluation Command
BTA	Business Transformation Agency
C&A	certification and accreditation
C4ISR	command, control, communications, computers, intelligence, surveillance & reconnaissance
CDR	critical design review
CHSS	commercial off-the-shelf hardware, software, and services
CIO	chief information officer
CJCSI	Chairman of the Joint Chiefs of Staff Instruction
COTS	commercial off-the-shelf

DAA	designated accrediting authority
DAMS	Defense Acquisition Management System
DAPA	Defense Acquisition Performance Assessment
DAS	Defense Acquisition System
DAU	Defense Acquisition University
DEP	distributed engineering plant
DIA	Defense Intelligence Agency
DIACAP	DOD Information Assurance Certification and Accreditation Process
DNS	Domain Name System
DODD	Department of Defense Directive
DODI	Department of Defense Instruction
DT&E	development, test, and evaluation
EA	evolutionary acquisition
ERP	enterprise resource planning
FDD	feature-driven development
FSO	Field Security Office
IA	information assurance
IA C&A	information assurance certification and accreditation
IID	iterative, incremental development
IOC	initial operating capability
IP	Internet Protocol
IPT	integrated product team
JCIDS	Joint Capabilities Integration and Development System
JITC	Joint Interoperability Test Command
JMETC	Joint Mission Environment Test Capability
JOTS	Joint Operation and Tactical System
KPP	key performance parameter
LRIP	limited rate initial production
MAIS	major automated information system(s)
MCRC	metrics collection and reporting capability
MDA	milestone decision authority
MDAP	major defense acquisition program
NAVSEA	Naval Sea Systems Command
NSA	National Security Agency

OIPT	overarching IPT
OMB	Office of Management and Budget
OPTEVFOR	Operational Testing and Evaluation Force
OSD	Office of the Secretary of Defense
OT&E	operational test and evaluation
PDR	preliminary design review
PKI	public key infrastructure
PMO	program management office
PMT	portfolio management team
POR	program of record
PPBES	Planning, Programming, Budgeting, and Execution System
RDECOM	Research, Development & Engineering Command
RDT&E	research, development, test, and evaluation
SDCI	software development and commercial off-the-shelf integration
SDLC	software development life cycle
SDREN	secret defense research and engineering
SFMPL	Submarine Force Mission Planning Library
SoSIL	Systems of Systems Integration Laboratory
T&E	test and evaluation
TD	technology development
TRA	technology readiness assessment
TRADOC	Training and Doctrine Command
TRL	technology readiness level
V&V	verification and validation
WIPT	working-level IPT
XP	extreme programming